LIVING WITH THE LAND:

Communities Restoring The Earth

Edited by
Christine Meyer & Faith Moosang

A project of the British Columbia
Environment & Development Working Group
and IDERA

The New Catalyst Bioregional Series

NEW SOCIETY PUBLISHERS
Philadelphia, PA **Gabriola Island, BC**

Canadian Cataloguing in Publication Data

Main entry under title:
Living With The Land
(The New Catalyst Bioregional Series; 4)
ISBN 1-55092-192-4 (bound); --ISBN 1-55092-193-2 (pbk.)

1. Economic development—Environmental aspects. 2. Environmental protection—Developing countries. 3. Human ecology—Developing countries.
I. Meyer, Christine (Christine Ann). II. Moosang, Faith. III Series.

HC79.E5L58 1992 333.7'15'091724 C92-091181-1

Canada ISBN: 1-55092-193-2 (Paperback)
Canada ISBN: 1-55092-192-4 (Hardback)
USA ISBN: 0-86571-251-4 (Paperback)
USA ISBN: 0-86571-250-6 (Hardback)

Cover design by Susan Mavor.

Book design and typesetting by *The New Catalyst/New Society Publishers*, Canada.
Printed in the United States of America on partially recycled paper by Capital City Press, Montpelier, Vermont.

To order directly from the publishers, please add $2.50 to the price for the first copy, 75 cents each additional copy (plus GST in Canada). Send check or money order to:
New Society Publishers,
4527 Springfield Avenue, Philadelphia PA, USA 19143
or
P.O.Box 189, Gabriola Island B.C., Canada V0R 1X0.

New Society Publishers is a project of the New Society Educational Foundation, a non-profit, tax-exempt public foundation in the U.S. and the Catalyst Education Society, a non-profit educational society in Canada. Opinions expressed in this book do not necessarily reflect positions of the New Society Educational Foundation, nor the Catalyst Education Society.

Living With The Land is the fourth of *The New Catalyst*'s Bioregional Series which is also available by subscription. Write: P.O. Box 189, Gabriola Island B.C., Canada V0R 1X0.

The New Catalyst Bioregional Series

The New Catalyst Bioregional Series was begun in 1990, the start of what some were calling "the turnaround decade" in recognition of the warning that humankind had ten years to turn around its present course, or risk such permanent damage to planet Earth that human life would likely become unviable. Unwilling to throw in the towel, *The New Catalyst*'s editorial collective took up the challenge of presenting, in new form, ideas and experiences that might radically influence the future.

As a tabloid, *The New Catalyst* magazine has been published quarterly since 1985. From the beginning, an important aim was to act as a catalyst among the diverse strands of the alternative movement—to break through the overly sharp dividing lines between environmentalists and aboriginal nations; peace activists and permaculturalists; feminists, food co-ops, city-reinhabitants and back-to-the-landers—to promote healthy dialogue among all these tendencies working for progressive change, for a new world. The emerging bioregional movement was itself a catalyst and umbrella for these groups, and so *The New Catalyst* became a bioregional journal for the northwest, consciously attempting to draw together the local efforts of people engaged in both resistance and renewal from as far apart as northern British Columbia, the Great Lakes, and the Ozark mountains, as well as the broader, more global thinking of key people from elsewhere in North America and around the world.

To broaden its readership, *The New Catalyst* changed format, the tabloid reorganized to include primarily material of regional importance, and distributed free, and the more enduring articles of relevance to a wider audience now published twice yearly in *The New Catalyst*'s Bioregional Series—a magazine in book form!

The Bioregional Series aims to inspire and stimulate the building of new, ecologically sustainable cultures and communities in their myriad facets through presenting a broad spectrum of concerns ranging from how we view the world and act within it, through efforts at restoring damaged ecosystems or greening the cities, to the raising of a new and hopeful generation. It is designed not for those content with merely saving what's left, but for those forward-looking folk with abundant energy for life, upon whom the future of Earth depends.

Living With The Land is the fourth volume in the series; others include *Turtle Talk: Voices For A Sustainable Future,* (No.1), *Green Business: Hope or Hoax?* (No.2), and *Putting Power In Its Place: Create Community Control!* (No.3).

*

The *Bioregional Series* and *The New Catalyst* magazine are available at a discount by subscription. Write for details to: *The New Catalyst,* P.O. Box 189, Gabriola Island, B.C., Canada V0R 1X0.

Table of Contents

Acknowledgments

This book would not have been possible without the generous financial support of the following organizations:

Canadian International Development Agency, Public Participation Program; Canadian Employment and Immigration (CEIC); CUSO British Columbia's Sustainable Natural Resources Committee; United Church of Canada, Van Deusen Fund; Environment and Development Support Program of the Canadian Environmental Network; and British Columbia Council for International Cooperation (BCCIC).

The British Columbia Environment and Development Working Group (BCEDWG) is a coalition of environment, international development, youth, union, and women's groups. The members of the coalition include the British Columbia Public Interest Research Group; British Columbia Environmental Network; Burns Bog Conservation Society; Canada World Youth, British Columbia Region; Canadian Youth Network for Asia Pacific Solidarity; Cariboo Chilcotin Tribal Council; Cariboo Horse Loggers Association; Co-Development Canada; CUSO British Columbia; Global Village, Nanaimo; HOPE International; International Development Education Resources Association; Kootenay Center for a Sustainable Future; Latin America Connexions; Oxfam Canada; Participants in Development Society; Save The Children Fund of British Columbia; Save Georgia Strait Alliance; Social Planning And Research Council of British Columbia; South Pacific People's Foundation; Turtle Island Earth Stewards/TIES International; Teaching Support Staff Union, Simon Fraser University; Unitarian Church; Unitarian Service Committee; United Nations Association; YMCA International; YWCA International Cooperation Program.

We'd like to make special mention of Shauna Sylvester for her unwavering support throughout this project. Other book committee members were Tyhson Banighen, Roxanne Cave, Leslie Kemp, Gus Polman, and Steve Stewart.

We'd especially like to thank all of the contributors for their words and patience. We recognize that the writing of these stories was not done in isolation. Special thanks go out to everyone who offered their thoughts and help to the writers. Some of these people and organizations are Jesus Balgos, Leticia Baluyan, Archie Chantyman, Circling Dawn Collective, Larry N. Cruz, Spud Durutti, Liza Galo, Bert Groenenberg, Herb Hammond, Susan Hammond, Ana Hardoy, International Institute for Environment and Development, Chief Roger Jimmie, Kluskus Band, Rodrigo (Jing Jing) LaChica, Marcial Lopez, Enrique Mendoza, José Jesús Mendoza, Beth Neilsen, Permaculture Trust of Botswana, Gary Sodden, Ulkatcho Band, Greg Utzig, Michael Walsh, and Stefan Wodicka.

This book began with long conversations and letters to the following people and organizations who connected us with the communities and assisted with background research: Michael Ableman, Partap Aggarwal, Tade Aina, Mary Alegretti, Lydia Alpizar, Felix O. Amubode, Anthony Anderson, Bob Anderson, Lesley Anderson, Jean Arnold, Kim Arnott, Alan Artabise, Andy Barham, Nafisa Barot, Pauline Birks, David Buckner, Sue Cameron, Adrienne Carr, Kathryn Cholette, Moffat Clarke, Beryl Clayton, Alma Monica de la Paz, Ecological Agriculture Projects, Dania Edwards, Professor Jimoh Omo Fadaka, Debbie Ferren, Margot Francis, Paul George, Ann Gillespie, Maya Gislason, Terry Glavin, Gustavo Gonzales, Claire Gram, Rob Gransom, Nalaka J. Gunawardene, D'arcy Harris, Joanne Hayward, Susan Holvenstot, Alfredo Jiminez, Mary Lindsay, Steve Lloyd, Sue McIntosh, Ian MacKenzie, Don MacLeay, Don Mallais, Esperenza Martinez, John Masai, Kevin Matthews, Martin McCann, Christina Mendoza, Betsy Moen, Pat Mooney, Eva Murray, Nigerian Environmental Study/Action Team, Annung Nur Rachmi, Rita Parikh, Janice Pearlman, David Peterson, Mike Pool, Sharon Rempel, Mark Roseland, Shirley Ross, Sanjit Roy, Gulzar Samji, Doreen Salwood, Paul Shay, Elizabeth Spalding, Ian Stein, Alan Stevens, Marjorie Stewart, Denis Udall, Reynaldo Ureta, and Bintoro Wisnu Prabawo.

Because many of the stories in this book came from communities where the first language is not English, the assistance of translators was essential. They are Jorge Aguilar, Elizabeth Fortes, David Huxley MacKinnon, Martha, Nedjo Rogers, and Sylvia.

A very special thanks for their time and effort goes to Delores Broten, Guy Dauncey, Laurie MacBride, and Gaokgabelwe Dorothy Ndaba.

When things were rushed we found we could always count on Peter Kerr to see us through logistical storms. Others who helped with logistics were Steve Bugnowitz, Steve Chan, Garth Fraser, Margie Manifold, Kirsten Smestead, and Frank Tester.

We are not the only editors of this book. Others who took the time to read and comment on the stories were Sage Birchwater, Annette Desmerais, Robert Fox, Jeremy Higham, Bill Horswill, Helen Lee, and Miriam Palacios.

We'd like to thank Christopher and Judith Plant for their editing skills and support throughout what sometimes seemed to be a never- ending process.

The following people offered assistance that ranged from drawing detailed diagrams to inspiring us with their visions and ideas: Brenda Beck, Jennifer Crawford, Eric Harris, Juan, Alex MacKenzie, David Murphy, Peter Padbury, Monica Pohlman, and Mutang Tu'o.

For legal advice we'd like to thank Ruth Beck, Margot LaCroix, Willem Meyer, Murray Mollard, Carol Rublack and Robert Shepherd.

The maps in this book were hand drawn by Jeannie Mikita. The cover was designed by Susan Mavor. Both women brought their immeasurable talents to the project with excellent results.

Finally, we'd like to thank the people who are a part of our community. Friends and family—too numerous to mention—but they know who they are.

*

Grateful acknowledgment is made for permission to use previously published material from the following sources:

"Building Community Organization: The History of a Squatter Settlement and its own Organizations in Buenos Aires," by Ana Hardoy and Jorge E. Hardoy with the collaboration of Richard Schusterman, *Environment and Urbanization*, Vol.III, No.2, October 1991.

Creating Successful Communities: A Guidebook to Growth Management Strategies, by Michael A. Mantell, the Conservation Foundation (1990; Island Press, Suite 300, 1718 Connecticut Avenue NW, Washington, D.C., U.S.A. 20009).

"De Campesino a Campesino," *Enlace* (July 1990; Centro de Intercambio Cultural y Tecnico, Apartado A-136, Managua, Nicaragua).

The Plant Breeders' Rights Act: What It Really Means..., by Ron Elliot, Genetic Resources for Our World (1989; GROW, Suite 750, Slater Street, Ottawa, Ontario, Canada K1P 6E2).

"The Sacred Land Rights of the Cordillera People," *Kabalikat: The Development Worker* (no. 2, March 1989; Council for People's Development, P.O. Box SM 359, Sta Mesa, Manila, Philippines).

Shattering, by Cary Fowler and Pat Mooney (1990; University of Arizona Press, 1230 North Park Avenue, Suite 102, Tuscon, Arizona, U.S.A. 85719).

"The (Un)making of a Fisheries Law," *Phildhrra Notes* (January-February 1991; Philippine Partnership for the Development of Human Resources in Rural Areas, 20 Jose Escaler Street, Loyola Heights, Quezon City, Philippines).

Foreword

Of Cargo-Cults and Conferences

Christopher Plant

The people who live next to an active volcano on the remote island of Tanna in the southwest Pacific are renowned for having invented a unique and colorful cargo cult. Visited briefly by United States forces during the second World War, they witnessed with some incredulity the arrival of huge ships and airplanes that disgorged jeeps, bulldozers and crates of canned food and other supplies on to their shores. Where before they had been largely ignorant of the industrialized world—with the exception of the occasional Christian missionary who failed to make significant inroads into their lives—these villagers now suddenly witnessed the existence of material plenty on a scale never previously imagined.

The U.S. forces were gone from the island as quickly and mysteriously as they had arrived, but not without leaving an indelible legacy in the minds of their hosts. The people played this out by making a thick red cross and erecting it in a grassy clearing in the middle of their village; they also fashioned pieces of wood into the shape of rifles. Holding the rifles in the proper military manner, they then formed themselves into a small squadron and marched back and forth across the clearing every now and then. They believed that, by performing this and other rituals in the prescribed manner, voluminous amounts of "cargo" would eventually arrive for them via the mouth of the nearby volcano.

Known as the Jon Frum movement, apparently after one of the U.S. servicemen who made the landing on the island, the cult has been regarded by most people from the industrialized world with a mixture of humor, condescension and scorn. Yet, in North America, the same people fervently believe that, by driving through rush-hour traffic each day and spending forty or more hours a week at work, the "cargo" of industrial society will miraculously appear for them at their own clearings—the shopping malls and paved parking lots scattered everywhere throughout suburbia. The widely-held belief in the "cult-of-the-mall," however, is regarded as neither primitive nor sadly bizarre; on the contrary, it is the epitome of modern civilization. Though both are cults of materialism, weighing them up in terms of their social and environmental impacts is instructive. The Jon Frum cargo cult does little, if any, harm to people or place, whereas the cult-of-the-mall reduces its players to wage slaves and is the engine that drives the destruction of the planet ....

Tanna is one of the southernmost islands of the archipelago of Vanuatu, formerly known as the New Hebrides. My visit to the Jon Frum people occurred in the late 1970s during a period when I was working for the Vanuuaku Pati, then struggling for independence and later to become the first government of a newly independent Melanesian nation. This visit has lingered in mind, emerging time and again as I try to make sense of a rapidly changing world.

Another cargo cult existed in the north of the New Hebrides at that time, and the foreign forces opposing political independence for the colony tried hard and quite successfully to co-opt both such movements into their camps. This reveals the vulnerability of the cargo cult as a phenomenon: lacking a sufficiently broad understanding of the geopolitical realities of the day, the cargo cultists of Melanesia may indeed be environmentally and socially harmless, but they are unlikely to be able to effect profound structural change. Their magic is not powerful, nor cosmopolitan, enough for modern times.

The Vanuuaku Pati, however, was different. Led by the first "globally" educated New Hebrideans, the party was fully cognizant of the international context in which the country existed. *Vanuuaku,* meaning "our land," was deliberately used as a rallying cry to unite the 80 distinct island cultures of the group. Participatory democracy at the community level ensured constant consensus over party policies and actions. From this strongly constructed base, the Vanuuaku Pati was able to succesfully challenge its colonial overlords and gain political independence—its aim being the locally controlled, regionally-based and culturally harmonious improvement of all the islanders' lives.

Significantly, neighboring Fiji, where I worked for four years, slid more smoothly into independent nationhood some ten years or more earlier but, less radicalized in the process, went the way of the development cargo cult—the cult-of-the-mall. Counselled by foreign development "experts," it gambled extensively on tourism to bring in foreign exchange dollars. But research in the mid-'70s demonstrated that a major part of the tourist industry was foreign-owned. This meant, for example, that Australian visitors got to the country on Australian ships and planes, travelled within Fiji on Australian bus lines, and booked into

Australian- and American-owned hotels for the duration of their stay. All but a few cents of the tourist dollar left the country to line the pockets of the industry's overseas owners...

*

As I write this in early 1992, international attention is focussing on the Earth Summit due to take place at Rio de Janeiro in June of this year. Billed as the most important event since the 1972 United Nations Conference on the Human Environment held in Stockholm, this U.N. Conference on Environment and Development (UNCED) promises a great deal—in a nutshell, to secure an environmentally sustainable future for all nations of the world. Under scrutiny, however, the mammoth UNCED circus reveals itself to be dangerously akin to a cargo cult.

Mathias Finger, from Geneva, has followed UNCED's evolution closely. He regards the pedigree of the event to be revealing. Following the Stockholm conference, the United Nations instituted the U.N. Environment Program (UNEP) to facilitate world-wide protection of the environment. A decade later, after severe criticism of UNEP's abilities in this regard, the idea of an independent World Commission on Environment & Development (WCED) arose at UNEP's ten year review conference. Eventually created in 1983, the commission was chaired by Norway's labor party leader, Gro Harlem Brundtland. Canadian businessman Maurice Strong, who was secretary-general of the Stockholm conference, was a member of this commission. The WCED held hearings around the world, coming up with its official report, *Our Common Future*—colloquially called the Brundtland Report—in 1987. Much-heralded by some, and attacked by others, this report was responsible for popularizing the term "sustainable development," and for promoting the idea that development could continue pretty much in the same way that it had done up to now, provided we were nice to the planet. When the United Nations decided to convene the UN Conference on Environment and Development in 1989, Maurice Strong emerged again as its secretary general, along with numerous others who had played similar roles before. In short, the same people are conducting Brazil '92 as orchestrated Stockholm '72 and Brundtland '82; and the events are remarkably similar in aims, process and likely outcomes that focus on aid, transfer of technology and the strengthening of existing institutions, such as the U.N. Environment Program. (See Mathias Finger, "UNCED—An Exercise in New Age Politics?" in Sachs W., Ed., *Conflicts in Global Ecology,* London: Zed Books, 1992.)

This might not be so bad if the Stockholm conference and the work of the WCED had been particularly effective. However, for the most part, this has not been the case. In the absence of more wide-reaching change, the transfer of aid and technology has only deepened the crisis of the poorer, southern nations of the world, and the UNEP has failed to live up to expectations. This does not auger well, especially if one also takes into account the two sets of United Nations sponsored negotiations that are under way parallel to the Earth Summit process, on climate change—which has reached a dead end—and biodiversity which seems little better advanced. An attempted convention on forests has been abandoned, largely because of the unwillingness of the industrialized nations of the North to agree to the same criteria of sustainability being attached to the cutting of *their* forests as they are demanding of the nations of the South.

Many commentators in fact believe that little in the way of concrete and spectacular results can be expected from the event. Decisions, after all, will be made by the *heads* of nation states, despite the concurrent Public or Global Forums where citizens and nongovernmental organizations (NGOs) can "dialogue" with world leaders. Some regard the wholesale incorporation of NGOs—representing women, youth, indigenous peoples, environmental groups, students, trade unions, business and industry—less as a sincere attempt to hear the voices of the people, and more as a way to increase the visibility of the UNCED process, whether these NGOs promote or protest the process. Others argue that the whole premise of working on the basis of a natural resource conservation and environmental protection agenda seems guaranteed to lead to the same North versus South, environment versus development, polarizations. The very definition of the problem—of environmental degradation as being separate from "development"—also obfuscates the search for genuine solutions. The destruction of ecosystems everywhere occurs as an inevitable by-product of development for profit—the cult-of-the-mall—as opposed to development for people and place. To focus on the "problems" of the environment in isolation is a delusion. Also, the ethos of the event is troubling: it appears to favor strongly the idea that the future of the planet is best taken care of by enlightened "planetary managers" entrusted with making decisions for the benefit of all. Finally, there is the real danger that, with all the media razzmatazz, UNCED will *appear* to have done something significant, when, in fact, actual conditions may change very little...

Skepticism of the Earth Summit breaking new ground is further warranted when one considers that there is no shortage of understanding exactly what the pressing problems of the world are, and there is even widespread agreement on this. But what is desperately short is *action.* Word magic and technological sleight of hand are not enough to counter the immense economic and political power of the transnational corporations which defy regulation at all levels and which have foisted a new, all-time low, business ethic of "global competition" and "comparative advantage" on a reluctant world; neither is good will enough to repair the damage done by the steady dismantling of national controls on trade through the GATT (General Agreement on Tarriffs and Trade) and free trade processes—both products of the transnationals, and both the direct cause of working conditions and pollution in Mexico, for instance, that are as shocking as the worst conditions produced by the industrial ghettoes of nineteenth century England.

As the Earthscan team from London, U.K., puts it, the world's forests, soils, water and atmosphere cannot be protected in some magical, isolated manner. The only way to prevent environmental degradation is to change the *causes* of that degradation—to ensure serious changes in national and international production and economic systems. Earthscan also points out that it is participatory democracy that has proven to be the best system for fairly allocating and managing environmental resources. Subsequently, the UNCED process needs to lead to the devolution of government and the empowerment of local people in their communities everywhere, and to evolve a new development paradigm "that tackles big problems by thinking small." (See *Defending The Future: A Guide to Sustainable Development*, by Johan Holmberg, Steven Bass and Lloyd Timberlake, London: Earthscan Publications, 1991.)

To change the world radically, the UNCED process may be necessary, but it is by no means sufficient. Certainly, international negotiations and agreements are an essential ingredient of any feasible plan for reversing the current course of destruction. And, certainly, the effect of gathering so much nongovernmental opinion in one place and at one time is bound to have far-reaching consequences in the long-term. But the dire need of the times is to challenge the status quo at a profound level, something that southern nations are pushing for, and northern nations are resisting. Like the Vanuaaku Pati in the New Hebrides a decade or so ago, the *people* of both North and South are essentially struggling for independence, but, this time around, it is independence from *trans*national colonialism. In terms of the urgency of the task, merely calling for dialogue, continuing with business pretty much as usual, and refusing to deal with the deep structural problems of the world today is tantamount to resorting to a cargo cult on the world stage.

*

In broad strokes, this sketch provides a backdrop—the international context—for this book. Far from fostering a cargo cult belief in world conferences and planetary managers, what the British Columbia Environment & Development Working Group and IDERA have done in collecting these stories is to bring the debate over environment and development back to earth. While issues of development are conventionally regarded as "out there," as only relevant for the "Third World," this book helps us realize that, as Barbara Deming puts it, "We are all part of one another." Our struggles for an ecologically sane kind of development are common and, hearing the voices gathered here from around the world, we are reminded of that which we share, rather than our differences; reminded of how we might help as friends and neighbors, rather than as mere voters of a nation state. So, while the Earth Summit and all its trappings might be discouragingly distant and inadequate at the global scale, this book reminds us that we can act positively and effectively at a more human scale—a scale that *actually* makes a difference.

Turning the Brundtland Report's "sustainable development" on its head, *Living with the Land* demonstrates the need to *develop sustainability* in a wide range of human affairs, from fishing and forestry, to seed-saving and improving the lives of the urban poor. Equally important, it is a record of this kind of evolution toward sustainability actually *taking place* in widely different communities. It restores our faith in the initiatives of grassroots communities working at the local level. Time and again, the stories in this book—whether from India or Brazil, Thailand or Canada—reaffirm the fact that, given very *little* help, communities can take care of both themselves and the future of the places they depend upon, if only the cult-of-the-mall can be kept at bay.

Tackling big problems by thinking small, paying attention to nature as the most reliable guide for action, having respect for local cultures, sculpting local economies to fit local ecosystems, and devolving power to the community level—these are also key elements of bioregional strategy here on the continent of Turtle Island. As part of *The New Catalyst*'s Bioregional Series, this collection extends the customary ambit of bioregionalism from being primarily northern-oriented to being more globally conscious, and cognizant of the countries of the South in particular where, in many places, earth-centered ways and self-reliant traditions that we on Turtle Island are trying to re-learn still exist, and are also being created anew from among the destruction wrought by the cult-of-the-mall.

Living with the Land emphasizes that North Americans can learn as much from the South, as the South can learn from the North. It says, too, that we can act here and now to transform this system from the grassroots, to get off the backs of cultures and communities in the South, and so contribute to bringing about a kind of change that has the power to be much more than a cargo cult.

By Word of Mouth: An Introduction

Christine Meyer and Faith Moosang

Land and *people* are two basic words that describe the contents of this book—two words that lie at the heart of many global, political, economic, and social tensions—two simple words that are becoming more difficult to understand. Why? Because our connections to each other and to the earth are disintegrating. We are reminded daily of the environmental crisis facing the planet; many communities are toppling as their resources are pulled from under them by the demands of international markets. Farmers everywhere are struggling to keep the soil on the ground, with the wind as their ghostly opponent.

In looking for solutions it has become more difficult to turn to the person down the road or next door, or even to one's own family. We are also facing a breakdown of community; in industrialized centers, air, water, and noise pollution take their toll on the physical health of the population while loneliness knocks on peoples' doors. Families are split apart as individuals migrate in search of jobs.

The picture seems bleak, but looking closer we find a brighter image growing in our midst—a picture of communities of people who are becoming reintegrated with their neighbors and the land around them.

It may be a surprise for many to learn that homeless people in Vancouver's inner city are finding a resting place in a community organic garden, or that farmers in Nicaragua and Indonesia are stemming the tide of soil erosion by teaching each other more appropriate farming techniques. Adding to this brighter image is a community of tribal people in the Philippines, who realized twenty years ago that gaining security of their land tenure and educating their young about the local area were the essential steps to ensuring a strong community. Meanwhile, displaced people squatting on land on the outskirts of Buenos Aires have formed their own representative organization to improve their quality of life. These are just some examples of the stories in this book that show how restoring the earth begins at the community level.

What's in a Word?

These communities tell us that the idea of "sustainable development" does not come from the pens of international policy-makers. In fact, many grassroots communities have known for centuries about the importance of linking concern for the environment to the way a community develops.

The term *sustainable development* was coined by the Brundtland Commission on Environment and Development and popularized in the book, *Our Common Future*. The commission attempted to address the problems caused by conventional development which has led to the exploitation of both people and resources in many areas. The addition of the word *sustainable* was a way to link environmental concerns to development. But the words do little if the very premise of development is not brought into question.

It has become commonplace to regard countries in the northern hemisphere as "developed" and those in the southern hemisphere as "developing" or "underdeveloped." This delineation suggests that development is synonymous with economic growth. The assumption is that if countries increase their rates of economic growth, there will be more national wealth that will trickle down to raise the living standards of everyone. This can be thought of as a "top-down" approach to development, where decisions are made primarily on the basis of international economics.

In the name of economic growth, dams are built, trees are cut down, and the concrete spreads. But who does this development serve? When market forces become the basic factor of development, local people and land are often left out of the equation. Even when the economic indicators imply that an area is developed, the reality may be the opposite. In most northern countries, where the gross national product (GNP) is high, rates of poverty, crime and homelessness have grown along with destruction of the environment.

When communities take control of their own development, from the bottom up, they can ensure that their basic rights are met. These rights, which include food, shelter, education, and human dignity, are not necessarily respected in conventional top-down development. Conventional development has also violated the rights of the earth. Seeing the destruction of forests, oceans, lakes, rivers, meadows, and neighborhoods around them, people are reawakening to the knowledge that they cannot live without recognizing the rights of the land.

Our coalition, the British Columbia Environment and Development Working Group (BCEDWG), is attempting to redefine the concept of development, rooting it in empowerment and community control. A community worker from India presented us with a challenge. He asked people in the North to show what they meant by sustainable development. He said he was tired of the words, theories, and discussions of development that are created by the North and exported to the South. He wanted to hear about what was working positively at the community level and why.

We decided to throw the challenge back to the grassroots. We asked, "How is your community managing to meet your needs and aspirations and those of future generations, while living within the ecological limits of your region and the earth? Tell us how you live and from that we'll learn about developing sustainability."

Our assumption in creating this book has been that even before the term *sustainable development* was coined, many communities already had generations of practice living with the land and not at its expense. We felt we could learn the most from these communities by listening directly to the voices of their members. We held back from imposing our idea of what we considered to be sustainable. Instead, we felt that these local communities were the best judges of whether or not their activities were promoting and exhibiting an ethic of dignity for themselves, their neighbors, and the earth.

Not knowing exactly what we would find, we contacted people as close as our local organic-food store and as far away as the Philippines and Argentina. We could only trust that somehow our letter requesting personal stories about grassroots initiatives would reach the people directly involved in their own community development.

The words in this book come from the people who are actually doing the work to restore the earth. They have an integrity that is largely missing from the writings of theorists and policy-makers because they come directly from experience. These are the words of action!

Actions Speak Louder Than Words

As the stories arrived on our desks, we realized that this was not only the beginning of a valuable lesson in how communities live with the land, but also a basis for people to connect and share their experiences with others.

These stories defy political borders and cultural boundaries with their similarities. In different words, they reveal the importance of knowing home, community, and region. They tell us that only by knowing one's home place is it possible to live in balance with the people and land there.

They also tell us of the importance of ensuring that this knowledge is passed to the children of the community. The thread of education links many of the stories, suggesting that the only way to create lasting change is to ensure that future generations will also care for the place they call home. This philosophy is reflected in the names of many communities' organizations. The Kalahan Education Foundation in the Philippines was formed by the Ikalahan tribal people in northern Luzon island. Organizing their community around the creation of a school ensures that young people do not have to go far to complete their regular schooling; at the same time, they can learn about the culture of their people and the ecological aspects of the land where they live. This kind of education helps to create a lasting commitment to community. If young people do go on to higher education, they are encouraged to return home and, as the Ikalahan people say, "learn to become good elders when it is their turn."

The Kluskus and Ulkatcho indigenous communities in central British Columbia, Canada, also see local education as one of the best ways to gain control over their land and resources. These communities have been struggling to contest logging in their traditional territories—logging that has not only eroded their livelihoods, but also their lives. The communities know that when the day comes that they are the ones responsible for the use of their forest, their children will need to learn not only about ecosystems in general but, as one chief puts it, "they will need to learn about the trees in *this* forest and the fish in *this* river."

Many communities also recognize that it is not just the children who need to learn more about their local area. Locally relevant education is the basis of the Foundation of Education for Life and Society (FELS), a hill-tribe community organization in northern Thailand. By visiting each other's farms, four villages are learning more about growing food in a way that does not harm the forest and soil around them. Deforestation not only threatens their ancient respect for the land but forces migration of young people to the cities, further eroding cultural traditions. FELS is working to encourage young people to reconnect with their traditional values through sharing between youths and elders.

Education between communities is also happening in Nicaragua with the Campesino a Campesino program. The success of this program relies heavily on the idea that farmers are more likely to learn from other farmers than from experts. Teaching is structured in such a way that farmers who are learning conservation techniques will be able to teach others, thus building solidarity among farmers while empowering individuals to take an active role in improving crops and income. Another example of this technique is seen in Indonesia, where most of the facilitators in the Yayasan Tananua program are farmers themselves. The goals are very similar—conservation and personal and community empowerment.

Restructuring the teacher-student relationship is also found in the work of MYRADA, an organization based in southern India. MYRADA learns directly from villagers and finds they often know more than the "experts."

> In one of the watersheds an area model was created and MYRADA staff asked the people to tell about the soils of the model. They ran to their fields, brought the soils and put them on the model. There were five types of soils. As MYRADA Executive Director Aloysius Fernandez relates, the staff later went to a university where experts pointed to a regional map asserting how, "in this area, they're all black soils, that area has all red soils, in that area it's all sandy loam..." But there were five types of soil in this one watershed. And when Fernandez asked the agriculturalists, "Which of the five would you think is the most productive?", they pointed to the wrong one. They didn't point to the one the people indicated. Then Fernandez said, "If you wanted to irrigate, which soil would you irrigate?" And again they were wrong.

A slightly different example is found in Canada. Thomas Evans works with urban and rural communities to encourage the planting and saving of diverse varieties of seed. He comments on how important it is for him that his workshops offer a venue where people can talk to each other and share what they know. The workshops bring people together in the same space and they take over from there.

Even though many communities are becoming more aware of their local needs, only so much can be done to protect a way of life without legal recognition that people own the land they live on. In Thailand, the government has threatened to relocate the hill- tribe villages in order to set the area aside as a national park. Although the communities have been there for hundreds of years, they do not have legal title to the land. Traditional concepts of ownership and sharing are being eroded by much larger forces external to the communities.

The need for land security is a lesson that is echoed throughout many stories in this book. Many indigenous people are offended by the need to seek a legal document to permit them to remain on land they have occupied for generations. But some communities, like the Ikalahan people, recognize they could do more to conserve their environment and culture by gaining legal title to their land.

The situation is more difficult for people who are displaced from their lands and forced to live in squatter settlements. Residents of Barrio San Jorge on the outskirts of Buenos Aires, Argentina, have had little support to improve their homes and surroundings. Until recently they have been living amidst garbage and pollution without the legal right to demand basic services. Despite this, and having been thrown together by circumstance, the residents of the barrio managed to form a representative community organization, working not only for security of the land they live on but also for improvements in the infrastructure and services.

Other cities have been more supportive of their low-income areas. To reduce garbage in Curitiba, Brazil, the city now encourages recycling in local neighborhoods. The neighborhoods arrange the collection of the recyclable materials, which the city takes in exchange for transportation vouchers or bags of surplus vegetables. The city planning department in Curitiba is particularly active in promoting a "green" city through intensive education and planning that encourages use of mass public transport. Although many of the initiatives come from the municipality, local communities also take responsibility for the greening of their own neighborhoods. The story of Curitiba is one example of how a large organization can work with communities to promote sustainability.

Many of the communities included in this book work with the help of outside groups or funders. The forces that break down communities and ecosystems have sometimes proven too powerful to be overcome by a single community acting on its own. Outside groups, often non-governmental organizations (NGOs), have acted as facilitators or advocates in the process of community empowerment. These groups also have lessons to share.

The Campesino a Campesino program in Nicaragua is funded by a number of international NGOs, but the actual work is facilitated by the National Union of Farmers and Ranchers (UNAG). A similar situation occurs in Indonesia, where the work of the Ecological Studies Project (ESP), a local NGO based in Solo, is funded by a Dutch organization. All the work is carried out by the communities in which ESP is helping villagers to create seed banks of local plant foods.

The role of outside organizations is examined honestly by members of the international NGO that is now working in the Barrio San Jorge. The frustration and cynicism of the squatters, who have seen many different organizations come and go from their community, should be a powerful reminder to outside agencies that they should work only on what community members identify as their needs.

Whether communities advocate the need for locally relevant education, security of their land, or saving traditional seeds, they share the same message: the people who commit to and understand the place where they live have the answers for restoring the land around them. It is up to others to listen to their words and provide the space or the contact that is needed to restore the earth.

Final Words

In creating this book we realized that no community stands in isolation. Many of the storytellers have expressed how, in their attempts to renew their land and spirit, they often feel like sparks in the wilderness. This book is a chance to create links so that these community sparks aren't swallowed up in the dark. It can be used as a creative tool to link with other communities, share ideas, and, finally, help us feel part of a larger community connecting people to land to people.

Part One

FOLKS IN THE FOREST

You ask us if we own the land and mock, "Where is your title?" When we query the meaning of your words you answer with taunting arrogance. "Where are the documents to prove that you own this land?" Title. Documents. Proof. Such arrogance to speak of owning the land, when you shall be owned by it. How can you own that which will outlive you? Only the race owns the land because only the race lasts forever.

—Macliing Dulag, Kalinga elder

Contacts:

Delbert Rice, Executive director
Leticia Baluyan, Project Coordinator (Health Program)
Moises Pindog, Project Coordinator (Agro-Forestry)

Kalahan Educational Foundation, Inc.
Imugan, Santa Fe
Nueva Vizcaya 3705
Philippines

4 Judge Luna Street
San Francisco Del Monte
Quezon City, Philippines

tel: (632) 98 10 92

1

Building Community on Ancestral Ground: The Ikalahan People

The Ikalahan Elders

Land and land rights are concerns common to indigenous communities worldwide. External forces continue to encroach upon ancestral lands. Whether this takes the form of industrial development or the usurping of lands by other peoples, the result is often cultural and environmental degradation. This occurs particularly when others take the land as their own without respecting and understanding the knowledge that comes from generations of ancestors.

Tribal communities in upland areas of the Philippines are facing the erosion of their traditional forest lands, threatened not only by logging but also by people from lowland societies who are searching for a source of livelihood. Often these new inhabitants do not share the knowledge and practice that has sustained traditional forest dwellers for generations. The situation has led tribal communities to find ways to protect their rights to land as well as the fragile ecosystems in which they live. The Ikalahan people have been doing this for more than twenty years. This is their story.

What follows is drawn from a composite statement of the Ikalahan elders, translated from the Ikalahan language by Delbert Rice and Leticia Baluyan.

We are the Ikalahan people who live in what is now known as the Kalahan Reserve—a 15,000-hectare area covering extensive forests and steep mountain lands. The forests serve as a watershed for much of northern Luzon Island, but they are also the source of our livelihood.

These mountains and some of the adjacent lowland areas are our ancestral home, but slowly the lowland peoples have taken away the lower portions from us and have been moving into the upland areas. This has made it very hard for us because our population is growing also. About twenty years ago some lowlanders were even able to get title to almost 200 hectares of our lands by bribing a judge. We had to go to court every month for more than two years and we had a lot of expenses. We finally won and their titles were canceled. The good thing is that this process put us onto the path toward getting control of our ancestral lands for ourselves. Actually, of course, getting control of the lands was only the first step toward protecting, preserving, and enhancing our resources and culture, all of which, together, will help us to improve our quality of life.

The elders had the vision—the whole community is making it a reality.

How We Gained Back Control

It is a complicated story but we will try to make it as clear as possible. First we tried to organize a producer's cooperative, which we were told would help us to market our products so that we would not be cheated quite so much. We spent a long time designing the by-laws and other papers so that the cooperative could really function in our own society with our own people. We, the elders, had regular meetings for over a year just to work out the structure of the cooperative, but when we finally went to the government office they would not allow us to register it. They wanted us to make some changes that would give them a lot of authority over our activities, even over our bank account. We did not trust them with that kind of power over us. We already had too much experience with things like that, so we just put our papers in the drawer and forgot about them.

We know that if we are to become self-sufficient, we must first protect and develop our natural resources.

By that time some of our young people wanted to go to high school. There were no schools in our mountains, so we had to send them to lowland schools. The problem was that after going to school in the lowlands, we found that they had changed and some had begun drinking too much alcohol. Why should we spend money for the "education" of our children if it just destroyed them? Finally we saw that there was another possibility, and so we asked Pastor and Mrs. Rice to open up a

Going to market: Children in the Kalahan Reserve learn all about the forest, including how to make useful items like brooms from leaves and branches. Photo: KEF.

high school for us in the mountains. They had been living with us for several years already and had promised to do whatever they could to help us. At first they were hesitant, but we convinced them. After a long study Pastor Rice found out that we would have to register as a legal organization if we were going to have our own school. We remembered the cooperative that we had tried to organize before. He dug the papers out of the drawer, changed the name from producer's cooperative to Educational Foundation, and made a few other small changes; we took the same papers to a different office to register them. This time we were successful.

With our new status as a legal organization, we were able to go to the government agency that administers the forest lands and finally, after several years, complete negotiations with them for control of almost 15,000 hectares of our ancestral lands. The contract we negotiated comes up for renewal every twenty-five years.

Our School

We wanted our school recognized by the government, so we found qualified teachers from other mountain tribes to teach. We did not want teachers from the lowlands because they do not seem to understand our culture. We told the teachers that they should teach the usual subjects, but that they should also teach children to respect our culture. Inherent in our culture is a respect for the land. We want children to learn about ecology and have set up a garden near the school, where they learn about organic growing. They also learn about how we protect the natural systems around Kalahan.

We also want the children to learn how to work. What good is a diploma in the pocket if the owner does not know how to earn a living? We want some of the graduates to go on to college, of course, to take advanced courses that we need locally, like forestry, engineering, education, food processing, and almost everything else, actually, except perhaps marine engineering or fisheries. There is not much need for those two when you live on steep slopes a kilometer above sea level! We do not want our young people to go away and work in other places because we need them here in the mountains.

How We Make Decisions

Each of the areas inside the reserve elects a representative, usually a tribal elder, to the Board of Trustees. The eleventh member is elected by the alumni of the academy. The board meets every month and usually spends all day discussing various issues about the land, school, and other problems that we have. At one time a bishop from our church said he too wanted to be a member of our Board of Trustees, but we politely told him that he was not qualified. We like our bishops but they are not from the mountains and do not understand our problems the way we do. These are *our* problems and we have to solve them ourselves.

We are self-governing, but not yet self-sufficient. Perhaps we could have been by now, but we had a very serious earthquake in 1990 that was followed by four equally serious typhoons. These really hurt us because all of our roads, trails, and water irrigation systems were completely destroyed. Many of our rice fields, some of our forests, and 80 percent of our homes were also destroyed, and more than twenty of our people were killed by these calamities. Some government people said we should abandon the area entirely, but where could we go? No place else seemed to have as much promise as our ancestral land, in spite of its problems. We have one advantage. We have already had our big earthquake. Every other part of the Philippines is in the earthquake zone, too, and if we moved somewhere else, we might just move into the area of the next big earthquake, so what's the use? We decided to stay.

As soon as the typhoons were finished, we carefully but quickly gathered available tree seedlings from our nurseries and used them to reforest the barren lands where slides had taken place. We planted more than 70 hectares before the rains stopped that year. The trees are already growing well and green slopes have replaced some of the brown patches that stained our mountains in 1990.

The Route to Self-Reliance

We know that if we are to become self-sufficient we must first protect and develop our natural resources. We hired some of our own young people who are now foresters and agriculturalists to work as a team to guide us in this. With their help we decide which trees to cut for building our houses and which areas to keep for permanent watershed and wildlife preserves. We don't do any agriculture in those preserves, but our animals play in them and we gather wild fruit and other minor forest products in them. Some of our staff are also working to find better varieties of sweet potatoes to plant in our *swidden* (slash and burn) farms.

Because our lands are all very steep, we cultivate each field for only two or three years before fallowing it. We usually leave the land fallow for fifteen years, but we recently discovered that when we plant new trees along with our sweet potatoes, the necessary fertility can be restored in less than eight years. We do not want to overcultivate our lands, because we know that would damage them. We are also discovering other methods for improving our food production and income while protecting our soils.

We have worked out a technique by which we can cut a few trees each year from each of the forest areas, to provide building materials for ourselves, raw materials for furniture construction, and a little lumber for sale to the lumberyards in the lowlands. They need wood too. We do all of this by hand so that the machinery will not damage the other young trees. This will provide the foundation with a little income without damaging the watersheds at all. The income may not be very much, but it will help.

Our staff invited teachers in to help us understand ecology. After several meetings, we decided that we would not allow any more chemicals, pesticides, or fertilizers into the area. This is not so hard to accomplish because we are all cooperating. Now some of our farmers are beginning to use their level lands to produce organic vegetables for the lowland market. We think it will be profitable and we know it will be better for our land. Some of our staff are even working to find better kinds of sweet potatoes for our basic food and to improve our ways of producing them if they can.

Generating an Income

We have to think seriously about our foundation income because we do not expect either our high school or our health center to become self-sufficient, and of course our office personnel do not produce any income at all. They will all have to be subsidized by the other projects of the foundation.

Some money will come from the many fruit trees we are planting. We can sell some of the fruit fresh when the trees begin producing well. Sometimes, however, the price goes down when it is in season, so we built a food-processing center where we turn those fruits into jams, jellies, and marmalades. We actually started the food-processing program to process wild fruits rather than domesticated ones. Now we are trying to domesticate some of the more valuable wild fruit from the forest in order to have a better supply.

Though it took us a long time to figure out how to market our food products, now they are being sold regularly in the best supermarkets in Manila and even in Europe. They have not yet produced a net profit, but we are hopeful that we will do so either this year or next. Eventually the profits from the food- processing program should subsidize the school. We collect tuition, of course, but not enough to cover all of our expenses.

Now we are planting some spice orchards. We plant mostly those that are hard to produce elsewhere. A climate such as ours is ideal for some spices. We have extra heat from the chimneys of our jelly factory, which we will use for drying the spices before packaging. We expect the income from this to subsidize the health center.

The carpentry and mechanics shop should be able to finance itself. We don't expect it to make a profit, but we hope that we don't have to subsidize it. We need the shop because the people there maintain our water and power systems and also manufacture many of our tools and most of our equipment.

Our Health Program

Our health program works more on prevention than on cures. That is why we call the building a health center rather than a clinic. Medicines are helpful when necessary, but a healthy lifestyle is much more important than medicines. The health center is only a part of the solution; the rest is up to the people. Our biggest job in the health program is to identify those who have tuberculosis or some other communicable disease and to get treatment for them. Next is getting vaccinations for all of the children. We do all of this through the community health workers. These people come from the surrounding villages into the center for training and return to their communities as volunteers in primary health care.

We know that our traditional supplies of protein are deficient so we are looking for alternatives. We are even using protein from leaves that are usually considered to be nonfoods, and we find it helpful. There are many other medicinal plants in our forests. We have already made a useful cough syrup that works well, and we think we have a root that will be very useful as a treatment for minor skin diseases. These are just a few more of the ways that we are trying to be self-sufficient.

Title to the Land

To go back to the original discussion about our lands, we will have to renew our contract in only a few more years. We hope we can change the contract into a communal title before that time, so that we will not need to go through the renewal process. Renewal should not be difficult for us, but none of us likes the principle involved. These are our lands and they contain the bones of our ancestors for many generations. Why should we have to get permission from anyone to use them? These papers are the best we can get at the present time, however, and they protect us from outside claimants and land grabbers. Some of our people are working with others in Manila to develop the means for us to get a communal title, which will probably be better. It has not been done before, but getting the papers for our lands had not been done before either. Now there are more than a dozen other tribal communities that have contracts like ours. Sometimes they come to our place to see how we do it. We are helping them and the government also.

Each family in our reserve actually owns their own lands. While we are too independent to do communal farming, our papers say that the land is owned communally. We like that because there are people in our community who might damage the resources by improper use. Because of the reserve, the elders have enough authority to make such people use the land properly. Communal ownership also ensures that the children and grandchildren can inherit the land, because no one can sell the land to outsiders.

These are our lands and they contain the bones of our ancestors for many generations. Why should we have to get permission from anyone to use them?

You see, we have had our own legal system in the mountains for hundreds of years. When we have problems, we call the concerned people together with our elders for a public meeting, and we spend as much time as necessary to find the truth and arrange for an effective reconciliation. After we have all the problems solved, we charge the guilty persons enough so that everyone can eat and dance together to wash away all the bad feelings. We do not wait for problems to become big or violent. We gather everyone together early so that we can *keep* the peace. The outsiders from the lowlands do not know how to do this anymore, so we have a hard time working together with them.

As soon as we got control of our lands, we made our policies on how we would protect and use the resources. At first we did not punish people very much if they violated the policies, but some people abused our patience. Now we are more strict. A few months ago, two men cut some trees in the watershed to make room for their new farms. We discovered it, of course, and made them reforest the entire area at their own expense and also fined them about one month's income each. That is only fair, because they were insulting the entire community that set the policies. They were also endangering the future of their own grandchildren.

We have much more contact with the lowlanders now and we have even started to dress like them. However, we still keep our culture for most things, because it is much better than lowland cultures. We are happy that our young people are studying in our own high school now, so they can learn to appreciate our culture better. Some of them will become good elders later, when it is their turn.

Contact:

Bert Groenenberg

Carrier Chilcotin Tribal Council,
301-383 Oliver Street
Williams Lake, British Columbia
Canada V2G 1M4

tel: (604) 398-7033

2

Regaining the Forest, Land, and Dignity: The Ulkatcho and Kluskus People

Bert Groenenberg

The Kluskus and Ulkatcho Indian bands in the central part of the province of British Columbia, Canada are part of the Southern Carrier Nation. For years they have had to deal with the effects of clear-cut logging encroaching on their lands. This kind of logging cuts all of the trees within a prescribed area, displacing humans and other animals. The Kluskus and Ulkatcho people have witnessed how clear-cutting has affected their environment and their lives. Now they want to bring the forest back into community hands. It's called wholistic forest use—an idea that combines traditional knowledge and uses of the forest with the realities of the local economy. This is the story of their struggle to create a more hopeful future for their children.

This story comes from observation of and conversations with Kluskus and Ulkatcho band members. Parts are taken directly from a case study written by Bert Groenenberg of the Carrier Chilcotin Tribal Council.

Standing among a grove of tall trees, Chief Roger Jimmie of the Kluskus people speaks slowly and thoughtfully. He makes a wide arc with his arm, revealing the hidden aspects of the forest around him.

He comments on the trees that have been knocked down by the wind. "We see this as good," he says. "They will break down and help the forest floor." He adds that the long, hair-like strands of lichen hanging from the branches of standing trees provide good food for the caribou. With a pocket knife he slices a thin strip of bark from a tree to reveal a white inner bark that tastes like a sweet fruit. He explains how a certain berry provides medicine and how the needles of some trees are used in spiritual ceremonies.

The chief also talks about the other uses of the forest—trapping for furs and selectively cutting trees. Previously, the band had much autonomy in these activities. They even began a small-scale milling operation to prepare the logs for building their own houses and schools.

Jimmie was elected chief at age 19, when there was a threat of a logging road being built into their traditional territories. The non English-speaking elders of the band decided that one of the younger English-speakers should be elected to help fight the clear-cuts. He is still chief today.

At a recent conference on sustainable forestry, he shared some of the philosophy that guides the community.

> Except for perhaps the fur trade with Europeans, we were pretty much independent and self-reliant in our roadless wilderness. We looked after ourselves. We could support large, extended families from the wealth of the forest. Moose, deer, squirrels were common animals we ate, along with mushrooms, roots, and berries. We also log, but only selectively. I can recall one of our elders saying, "Maybe squirrel, he climb that tree; maybe lynx use that tree over there." In other words, we respected all users of the forest. We were part of this forest.

The Encroaching Road

The problems for the Kluskus and their neighbors, the Nazko people, began in the early 1970s when a logging road was built west from the city of Quesnel into their traditional territories. Trap lines were destroyed along with other traditional uses of the forest.

The road went in despite blockades and protests, and soon the two communities felt the social consequences. People could no longer provide for themselves, and by the late 1980s most families were largely reliant on government financial assistance. This led to a pattern of violence and death. Within six months of the road's completion, six of the brightest young people in Nazko died of suicide or alcohol related incidences. Today, over 80 percent of band members depend on welfare. Some young people have known little else.

The band attempted several times in those years to suggest alternatives to the provincial government. The elders said that some logging could occur in those areas that were of low or minimal value to traditional uses. Selective logging would allow coexistence of both traditional and industrial forest use. But discussions with forestry officials and the

Learning From The Elders: At the Ulkatcho gathering, Sophie Thomas explains the medicinal qualities of the juniper branch under the watchful eyes of Curtis Leon.

Photo: Sage Birchwater.

logging companies proved frustrating and achieved little. At one point, the band managed to get an agreement for a ten-year moratorium on logging, but this was overturned six weeks later.

Kluskus people also tried to rid themselves of their dependence on money from the federal government. Band members felt that government aid created dependency and the loss of self-respect. In the early 1980s, along with other members of the Union of British Columbia Indian Chiefs, they rejected federal funding and retreated into their home territory. During this time they hunted and traded furs for cash to buy the provisions they could not supply themselves. For a few years they were essentially self- reliant. But the clear-cuts were coming closer and destroying more and more trap lines, until the community could no longer deal with the deaths and despair.

They did not give up. Knowing that an economy cannot depend on one resource alone, they had to find a solution that considered *all* aspects of the forest, including the people living there and dying because of its destruction. "What keeps us going," says Chief Roger Jimmie, "is we don't like to see our people dying left and right. That's what happens when other people manage your life, your economy, and do your logging for you."

Finding Common Ground

The neighboring Ulkatcho band joined forces with the Kluskus to find a solution. They hired forestry consultant Herb Hammond of Silva Ecosystem Consultants, Ltd., to help them come up with a plan that would take into account all the forest values.

They made an inventory of the forest by mapping traditional use areas. Using clear plastic overlays on topographical maps, the two bands began to outline moose and deer pastures, areas with different soils and trees, prime mushroom-growing and medicine-gathering areas. They identified good sites for tourist lodges and other recreational opportunities, as well as traditional family trapping and hunting areas. They even considered particular areas for timber harvesting that would have little or no adverse impact on existing uses. Lastly, they also looked at alternative logging practices. Logging with cable systems, horses, or smaller machinery would, it was hoped, prevent the kind of problems experienced with conventional clear-cut logging.

"We don't like to see our people dying left and right. That's what happens when other people manage your life, your economy, and do your logging for you."

The bands saw the need to work with other resource users in the area. The Ulkatcho band joined forces with those having similar problems with the large clear-cuts. This issue formed the basis for discussions between trappers, ranchers, guide-outfitters and lodge operators—between native and non-native people.

What eventually emerged was the West Chilcotin Resource Board, elected by each sector of the community in the summer of 1990. Now, when the Forest Service asks to meet with the Ulkatcho band, it is faced with a whole community, all opposed to existing clear-cut logging practices. The community is taking power back into local hands.

As Chief Roger Jimmie of the Kluskus said, "A few young people may be able to survive on logging, but the others need the hunting and fishing. They know about the waterways and animals. We had to develop a logging plan to accommodate that. The government says logging will happen anyway, so we had to come up with something agreeable to everyone, including the white people in the area, the rancher, the lodge owners, and the people who fish."

"You don't have to go to school to know 1000-acre clearcuts are bad."

After the communities had come up with a plan for wholistic forest use, they needed to find a way to gain back control of the forests. They decided to apply for a Tree Farm License from the government's Ministry of Forests. It was the best option because the license is based on a given area of forest land and not on the amount of wood taken out. Both bands applied for a Tree Farm License to cover much of their traditional territory and therefore protect some of their jurisdiction.

Frustrations and Negotiations

Unfortunately, the government had recently placed a ban on issuing tree farm licenses because of public protest about very large companies getting such licences. Although the bands had proposed to manage the forest in a sustainable way, the Ministry of Forests has refused even to consider their application because of the moratorium.

The government said it was still willing to listen to alternatives, but in discussions it became clear that ministry officials questioned the ability of the bands to do any kind of forestry management. The late Ulkatcho Chief Jimmie Stillas was asked how many band members had formal forestry educations. He replied, "You don't have to go to school to know 1000-acre clearcuts are bad." (In fact, some of the existing clear-cuts in the area are more than 2500 acres in size.)

As the negotiating continued, it became more evident that the government and the communities did not understand each other. The government recently proposed that the Kluskus band could cut 45,000 cubic meters of wood annually—an offer of such magnitude clearly misses the point of wholistic forest use, which considers the whole forest ecosystem. This difference in understanding has been one of the fundamental difficulties in the negotiating process.

Visioning the Future

Although the outcome is still uncertain, the process of fighting the clear-cuts has helped the bands to define a vision and strategy for the future. This vision is geared to building their self-reliance and looking at ways to combat internally the problems that have been created by the external management of resources.

The Ulkatcho band has developed a major initiative called the Ulkatcho 2000 Development Strategy, which states a vision for the future on all fronts. Through this plan they created the Ulkatcho Education Center which attempts to provide nonformal education opportunities more relevant to the local economy and region. Chief Roger Jimmie of the Kluskus band says he sees education as the most important step toward self-sufficiency. "The children need to learn about *this* area, *this* forest, and the fish in *this* river," he says. Locally relevant education will lead to more band-centered careers, perhaps in forestry or wildlife management.

The bands are also looking at ways to add local value to the timber that is harvested, before it is exported from the territory or its nearby urban center. They have ideas on how to sell lumber locally and how to

Wholistic Forest Use: What It Really Means

Herb Hammond

Wholistic forest use means the wise use and protection of forests throughout the full spectrum of human interactions with them. Forests are diverse, interconnected webs that sustain the whole (all life forms), not just one part (for example, timber). All parts of a forest have important purposes and must be respected and protected. People are part of forests. Forests provide air and water, moderate our climate, furnish homes and food for fish and wildlife, as well as people; provide spiritual renewal for all living organisms, and are an important part of human economies. When people change one part of the forest, all parts are affected. Thus, we must interact with forests in a careful, caring manner that protects all forest functions and components.

Wholistic forest use sustains human communities. This approach starts with a complete, field-based inventory of natural, social, and economic factors. From this inventory a wholistic forest use plan is prepared. The plan gives priority to the protection of natural factors, recognizing that healthy ecosystems are the bases for healthy economies and societies. Forests sustain us, we do not sustain develop secondary industries such as furniture making, paneling and log-house construction. This will bring more money to the local economy.

But it's hard to imagine this future while the negative effects of clear-cut logging are still so apparent in the community. The bands are not just fighting for their environment, but for their own survival.

Chief Roger Jimmie often remarks about the ironies of what is legal and illegal in our society. Half joking and half serious, he talks about taking more radical action, like hammering spikes into the trees. A civil disobedience practice known as *tree spiking,* it is done to make the lumber less valuable, so that forest companies will be more hesitant to cut the trees in the area. "The government has made spiking trees illegal," he says. "Maybe logging without looking at what it does to the people in the area should also be illegal."

forests. Providing a diversity of uses that do not degrade forests will protect them for our long-term benefit. Current "integrated forest management plans" do not usually meet these criteria and are timber biased, placing short-term monetary profits from timber extraction ahead of more balanced, ecologically responsible patterns of forest use.

Wholistic forest use proposes that we zone forests for a variety of activities on a watershed-by-watershed basis. Timber extraction is only considered as one possible forest use and is recognized as having potentially large negative impacts on the future growth of trees and on other forest uses such as fish and wildlife habitats, trapping, and tourism. Wholistic forest use recommends that we abandon harsh practices such as clear-cutting, slash-burning, pesticide application, and high-grading (removal of only the best timber). Where timber extraction is determined to be an acceptable use of the forest, selective logging methods are used and all logged trees are utilized, regardless of quality or size. Substantial parts of a forest, usually entire watersheds, are zoned or planned for nontimber uses such as cultural and spiritual places, water supply, fish and wildlife needs, soil protection, trapping, and various forms of tourism. Some of these uses have dollar values, others do not. All of these uses are vital for human needs and for the survival of the earth.

(Herb Hammond is a Registered Professional Forester. For more information about wholistic forest use, contact him at Silva Ecosystem Consultants, RR#1, Winlaw, B.C., Canada V0G 2J0; tel: (604) 226-7770, fax: (604) 226-7446.)

Contact:

Doug Gook

Cariboo Horse Loggers Association
P.O. Box 4321
Quesnel, British Columbia
Canada V2J 3J3

tel: (604) 747-3363

3

Stepping Lightly with Heavy Hooves: Horse Logging in British Columbia

Doug Gook

To walk lightly on the planet is one of those challenges that nurtures both our inner quest for harmony and the need to consider future generations. Sensitive selection-logging, with only the footsteps of horse and human left behind in the forest, may well be one of those options we need to encourage more.

The Cariboo Horse Loggers Association was formed in 1984 to support a more sustainable approach to forest management and community employment needs in the Quesnel area of British Columbia, Canada. The members promote a logging system that removes trees from the forest with minimal damage to what remains. They believe there is also a need to reverse the industrial forestry trend, whereby there is less employment for the community even though greater amounts of timber are cut. Selection logging with horses promises to provide a practical option that can satisfy these two major needs.

Doug Gook is a member of the Cariboo Horse Loggers Association.

I thought of horse logging for the first time eight years ago, when I was considering how I could best live off our small farm. Farming is fine when there is no snow on the ground, but I needed something for the winter. The farm had a woodlot and I began looking at horses to help me selectively log the trees there.

In 1984 I joined with other local people interested in horse logging on a larger scale. We figured that together, we would have a stronger voice to promote selective horse-logging and a better chance of securing contracts from the forest ministry. We called ourselves the Cariboo Horse

Loggers Association.

Slowly, we built up a wide variety of examples of forest areas that really benefited from our work. We were able to show the viability of selection logging in areas that the industry had never before considered. That was threatening to some in the industry; but we were a hardy group. We persevered and built up respect from individuals in the forestry sector.

The underlying assumption of our work is that natural forests know how to manage themselves.

Our association felt that in order to survive, we were going to have to provide a forest use system that was unique. We didn't want to be competing with machines. We felt that being able to maneuver timber out of a stand with minimal damage to the soil and remaining trees is what makes us unique. Horses are very adept at this kind of maneuvering. In addition, we held conferences, workshops, and meetings to develop a variety of standards and guidelines that we try to promote among all our members.

The underlying assumption of our work is that natural forests know how to manage themselves. The natural system has it worked out pretty well. Forests have been around for a few thousand years and have produced incredibly diverse and balanced systems and high quality wood. We cannot really improve a whole lot on that. We feel that working closer to those natural functions and processes is the safest way to go. We learn about wild-forest systems, how fire affects them, how bugs affect them, how different tree species affect them, and we do this for every site we work.

We also recognize that there is a whole array of other values in forests that go beyond wood and board. We are not just horse loggers; several of us are also involved in wildcraft. This means we enjoy going into the forest to pick mushrooms, medicinal herbs, and berries. We think it's important to remember that there are many reasons why we should just leave forests alone.

We have always said that in the horse-logging method, wild forests set the standard. We just have to drop a lot of our human ego and sense of domineering over nature. If we can lose that baggage, then we can see how forests have evolved their own sustainability.

Along with the ecological aspect, our association is also addressing the needs of people and livelihoods. Horse logging is extremely labor-

Skidding Through The Forest: Cariboo horse logger, Rob Borsato, with Laura.

Photo: Myron Kozak.

intensive, and that is the big difference between it and the mechanized systems used in clear-cutting. Conventional approaches are more capital-intensive, and so much of the resource money leaves the community to pay for the bank loan, to buy the machine, and to purchase the fuel and maintenance. But it's not just the cost of the machines that pulls money away from the community. The forest industry operates with very little respect for local communities because it works in a centralized, corporate, global economic structure.

Horse logging is one of those soft technologies that considers the needs of communities. It gives us more room to use less, to forego ever-expanding growth in terms of how much we cut, how much profit we make. I think that is going to be one of the keys to evolving a sustainable culture in this area. But it will only happen if the people recognize their right as the actual owners of the forest land in British

Just Add More People: "We figured that together we'd have a stronger voice to promote selective horse logging."—The Cariboo Horse Loggers.
Photo: Norman Jacob.

Columbia and demand alternatives to our current forest practices.

Horse logging is one of those soft technologies that considers the needs of communities. It gives us more room to use less ....

There have been high and low points for the Cariboo Horse Loggers Association. We have had three conferences, with people coming from all across Canada and the Pacific Northwest. But the big problem has been not having enough work. It has been hard to break into the process, to acquire the support and recognition to secure contracts. We have also had to reject some prospects because of our standards and guidelines. From past experience we knew that we would be at risk of compromising ourselves if we accepted particular contract conditions. For example, sometimes the forest ministry required us to take out many more trees than we thought was appropriate for the area. With horse logging you don't have to take out much at all to make it economical. We think it is important to realize that our best security as an organization comes from firmly maintaining our standards and quality of work.

One of the other difficulties we've encountered has been fluctuations in our membership. Currently there are only fifteen of us. We have had as many as thirty-five active loggers and enough work for six months to a year ahead. Now, quite a few people have left the association because there is not enough work. If there is a major shift in support and the number of areas available for us to log selectively, we may offer some training programs for serious people who want to make a go of it. We have been reluctant to sponsor more training projects because there is little point in training people to sit on the sidelines and watch all of these areas being clear-cut.

We have also had to balance individual interests within the group. Some people have rejected wholistic forest use in favor of making a bigger buck. But generally people have realized that in order for our children and future generations to have the opportunities that we have had, we are going to have to get more from less and do a better job all around in our use of the forest.

Contact:

Pedro Penafiel, Executive Director

Flor Figueroa, Legal Representative

Richard Murcia, President

Amigos de la Naturaleza de Mindo
Manosca 1011
Residencias Altamira
Casa 07
Quito, Ecuador

4

From a House without Walls: Friends of Nature, Friends of Life

Ramon Silva

Rather than arguing about the best ways to cut the trees, many people are learning that there are times just to leave the forest alone. Forest reserve lands are being created all over the world in recognition that forests have a value outside of human use. As long as local people are not forced to move from these areas, conserving forests may be a viable way to stem the effects of deforestation. Reserve areas can also educate people about nature. Until people rethink their relationship with natural systems, creating reserves might be the only way to save ourselves from ourselves.

Ramon Silva has been involved with Latin American grassroots groups both in Latin America and at his home in Vancouver, British Columbia. During his most recent trip to Ecuador in 1991, he met with a group involved in preserving their local forest.

During a recent visit to Ecuador, I spent most of my time with Amigos de la Naturaleza de Mindo (ANM), which roughly translates to "Friends of Nature of Mindo." I became interested in Mindo after a lengthy conversation with Pedro Penafiel, the executive director of ANM. Pedro and his wife, Cecilia, are active members of the community of Mindo. Together with their children they have built a log house without walls or windows, entirely open to its forest surroundings. They live next to a river where they collect their cooking and washing water. They have built a compost for all their organic leftovers. No plastics, chemicals, or other nonbiodegradable items come into or leave the house. Through their actions, Pedro, Cecilia, and their children are hoping to encourage other people in Mindo to adapt their lifestyles to their surroundings as a way of helping to protect the environment. The following is the story of ANM and the people of the community of Mindo, as told by Pedro Penafiel. When Pedro talks about his community, he begins by describing the beauty and tranquility of Mindo. He tells of its proud inhabitants and their dedication to hard work.

*

Traditionally Mindo remained isolated from the main urban centers of the country. The only form of communication was the small footpaths used by traders and farmers in the area. Because of this, Mindo remained virtually untouched by the outside. The people enjoyed its natural beauty, and all resources the earth shared with them. But, like many other growing communities, Mindo became more accessible to the outside world, and many people began coming to our community. People were looking for land to help feed their families. As more people came, more of our forest was cut down, and continues to be cut down to this day. We began to see the impact this was having on our environment and decided to put our thoughts into action. We needed to make people more aware of how fragile the forests really are, and of how easy it is to damage them permanently. We didn't want to lose the beauty of our community and have Mindo turn into another industrialized town.

More and more we recognized the beauty of our trees, the rivers, the birds, and all the other animals that live in the forests. We wanted to understand how they are all linked together and how we are linked to them.

In 1985, a group of us began to discuss the possibility of our forest disappearing forever. We saw this possibility as an important reason to begin educating ourselves and other people in the community about the environment. More and more we recognized the beauty of our trees, the rivers, the birds, and all the other animals that live in the forests. We wanted to understand how they are all linked together and how we are linked to them. After a while, more people started to become concerned about the destruction of the forest and how this was affecting their lives and those of their children. The following year, on October 10, 1986, we founded Amigos de la Naturaleza de Mindo. With the support of the German Technical Cooperation Service, we organized several social and cultural activities related to making our community more aware. In 1987 we began the Protected Forest Campaign and asked for the help of well-established organizations in Ecuador to assist us in our efforts.

At first we were very excited about our new organization and all the energy we had, but we soon found out it wasn't that easy promoting our Protected Forest Campaign. We encountered many obstacles. Our most serious problem was the fact that not all the members of the community supported our efforts to conserve the Mindo-Nambillo Forest Reserve and to stop its destruction. One example of this was that people who had recently immigrated to the area insisted on using the reserve and its natural resources for their own personal economic benefit. These people started to invade the reserve boundaries and set up farms. They began to exploit the forest for timber and other resources. They began to claim land, piece by piece, without really thinking about how their actions were putting the forest in danger. Although this threat to the reserve has decreased in the last few years, we feel that we must continue our efforts to educate the community about how to use resources properly without exploiting the earth.

Another problem we faced was our struggle to become a recognized organization within Ecuador. In our country, all civil organizations that are not registered with the government are considered illegal by the Ministry of Agriculture, which is responsible for land disputes and property and land protection rights. Because of this, we were seen as an irresponsible group, and they assumed we would not be able to carry out our goals. We suffered a lot because of this, and it was hard for us to feel good about our group. When we had problems trying to maintain areas of the Mindo-Nambillo Forest Reserve, the ministry told us that we should try to contact legal organizations within the country. In this way we became involved with Tierra Viva Quito, a leading environmental group. We began to get more support and felt more secure about going ahead with our objectives. Now, each month we send reports to the Ministry of Agriculture, and when there are serious problems we send in special reports.

Even before the ministry declared the area a reserve, members of the community of Mindo worked as volunteer rangers to make sure that remote parts of the forest were not being illegally exploited. Sometimes we were able to cover more than 35 kilometers in a day. This would lead us to remote areas and other small communities that border the reserve. Not only did we make sure that the reserve's resources were not being exploited, we also made a point of talking to community leaders and farmers. We discussed conserving the forest and asked for ideas and help. Since we have been collaborating with Tierra Viva and now have funding for our preservation project, we are able to pay these rangers a

Sign Of The Times: "The Mindo-Nambillo Protected Forest—the destruction of the vegetation, the wildlife and fish is prohibited ...."
Photo: Ramon Silva.

small wage for their time in the forest.

As a result of our communication with Tierra Viva, we together initiated the Conservation and Environmental Education Project on June 10, 1989. Tierra Viva was able to obtain a three-year grant from the World Wildlife Fund to support our joint project. Since we began the project, we have developed a program for the schoolchildren in Mindo and neighboring villages. These projects really began in 1986, but starting in 1987 we developed new programs for Mindo's young people during the three months of vacation—February, March, and April. We also organized events for the community such as seminars, courses, and festive activities. Since we started this we have had a bit more money with which to put together a more interesting program. A part-time coordinator now works every week or two, carrying out activities such as arts and crafts, community events, and theater presentations.

Another part of the project was the construction of an environmental learning center in the protected forest. We spent many months building the center, always considering how it fit into its surroundings. Since Cecilia and I already had experience building our house, we had a pretty good idea of how the center should look. Now that it's built, many people stay there and learn about conserving the forest. We encourage students and environmentalists as well as national and international tourists to come and experience our center. Some people come only to enjoy their holidays, others come to learn about the plants and animals around the center, but everybody comes because of the beauty of the forest and the sounds of all its animals. We are hoping that the money we collect from admissions to the center will eventually cover the costs of maintaining its facilities and the trails we've built around the area.

The Mindo-Nambillo Forest Reserve

Ramon Silva

In Ecuador there is a strong and active environmental movement. Several different groups share a common goal—to prevent the further degradation of Ecuador's environment. The interaction of these groups has led to the formation of a coalition of environment and human rights groups. The coalition's primary objective is the conservation of natural protected areas and the defense of the indigenous peoples of Ecuador's Amazon basin.

The group Amigos de la Naturaleza de Mindo (ANM) has been particularly involved in the activities of the coalition and works actively with the community of Mindo to promote environmental awareness. Their work centers around the upgrading and protection of the Mindo-Nambillo Forest Reserve, 19,200 hectares of mountainous topography bordering the community of Mindo. The reserve's lowest altitude is 1400 meters, while its highest points are found at the summits of Padre Encantado (4750 meters) and Guagua Pichincha (4780 meters).

Within Ecuador's system of wilderness areas, there are fifteen reserves that have a protected status similar to the Mindo-Nambillo reserve. The conservation of Ecuador's northeastern mountain forests is of vital importance because of the increasing pressure from both local communities as well as commercial development interests. In the case of the Mindo-Nambillo reserve, this pressure has been heightened by the opening of a new road on the reserve's perimeter. ANM collaborated with another organization, Tierra Viva-Quito, to upgrade the Mindo-Nambillo reserve to the status of protected forest. Their persistent campaign paid off on April 12, 1988, when the reserve was declared Bosque y Vegetacion Protectores de Montanas y Cordilleras de Nambillo (Protected Forest and Vegetation of the Mountain Chains of Nambillo) by Ecuador's Ministry of Agriculture. There are no communities inside the actual reserve, although there are five privately owned lots. These individuals have accepted the terms of the accord with the ministry, which limits their activities within the reserve.

We have many new ideas for the future and would like to make some new contacts with organizations in other countries. For us the most important thing is sharing information about the use of resources while helping to conserve them for future generations. We have started to prepare an agro-forestry project with a few courses in organic agriculture. Soon we want to begin construction of a tree nursery. Right now we are looking for funding to get moving on this project.

Some people come only to enjoy their holidays, others come to learn about the plants and animals around the center, but everybody comes because of the beauty of the forest and the sounds of all its animals.

Our long-term goal is to educate ourselves more about renewable resources. In other words, if we want to cut down trees selectively, we must also plant trees to replace the ones we cut down. Planting for harvesting can be a good way for all the farmers in the area to make a living while giving back what they take from their environment. I think this is very realistic for Mindo and other communities close by, because people here are used to working the land to make their living. In this way we can make sure that our children and grandchildren will have the resources they need for the future. If we can accomplish these goals, some day it may not be necessary to go into the forest for the wood to build our houses and earn our livings.

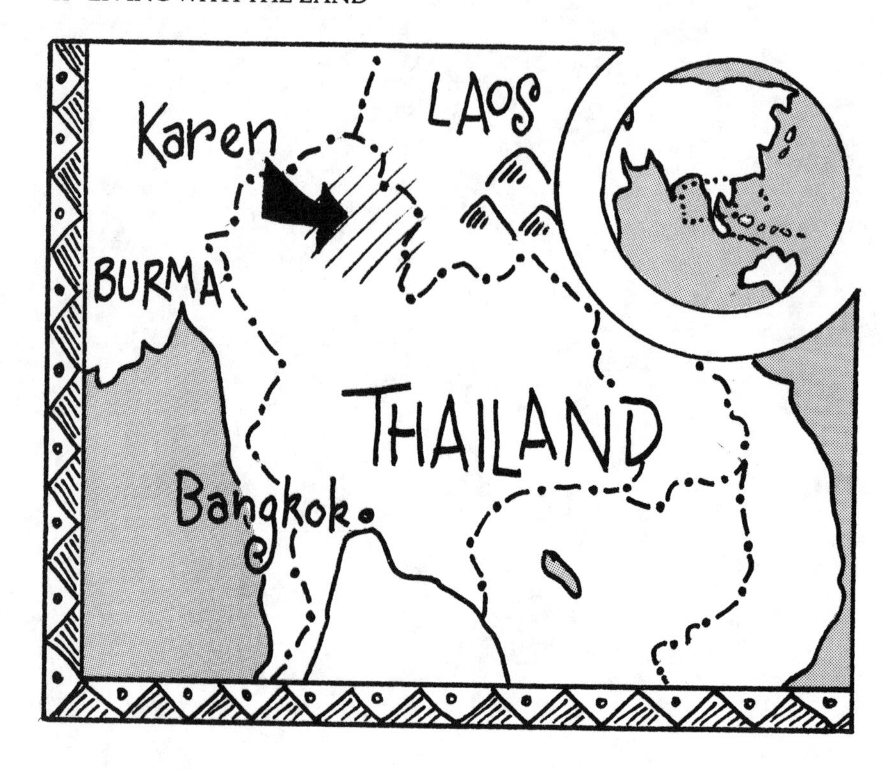

Contact:

Ms. Busaba Paratacharya

Foundation of Education for Life and Society
47 Phaholyothin Golf Village
Phaholyothin Road
Bangkhen, Bangkok 10900
Thailand

tel: (66) (2) 513-3038 and 513-4408

5

Education for Life: The Hill Tribe Development Project

Busaba Paratacharya and Suphan Inphu

In Thailand there are those who think the activities of tribal people are the cause of deforestation and watershed degeneration. This sometimes leads to policies that advocate removing all tribal people from the hill areas or pressuring them into becoming cash-crop farmers. There are others, however, who see that it is important for indigenous people to remain on lands they have occupied for generations, evolving self-sufficient communities in harmony with nature.

The Foundation of Education for Life and Society (FELS) is a nongovernmental organization that has been working in rural development in Thailand for the past twenty years. At present there are twenty staff working in four community development projects with lowland farmers in the north and northeast regions and with tribal people in the north. FELS believes in the principles of self-reliance, sustainable development, and participating with tribal communities.

FELS manager Busaba Paratacharya spoke with Suphan Inphu, the coordinator of the Hill Tribe Development Project. This story is translated from that discussion.

I was raised in a northern Thai village and completed my schooling in the north as well. I have always been keenly concerned with the situation of the hill-tribe people, and rather than move into an established city job, I decided to work as a walking teacher with the Hill Area Education Project, working first with the Yao tribe and then the Karen tribe. As community development is part of a walking teacher's role, I was eager to volunteer to assist FELS when they became interested in the Karen villages in my area. After almost five years I have moved on to become project coordinator for the Hill Tribe Development Project, which centers on the four Karen villages where I used to teach. With me are three volunteers, two northern Thai and one Karen.

The four villages where we work are in the hills of Lampang Province. These forested hills comprise the watershed of three main rivers and in the past provided 417 villagers in 87 households with water year round. As in my teaching days, I still enjoy walking through the rich forest between the villages.

Today, though, we are facing a large problem. The government has earmarked this area as a national park watershed reserve. I fear they will want to relocate the villages. Even though the Karen have been here for several hundred years, they have no legal title to any land. Despite this they are working hard in their communities to combat the more recent problem of deforestation along with a flood-drought cycle creating seasonal water shortages.

Even though the Karen have been here for several hundred years, they have no legal title to any land.

The consumer society of the lowlands is penetrating and damaging the Karen people's way of life. Once plentiful, "forest food" such as fish, shrimp, bamboo shoots, various tubers, and wild greens have become harder to find. The need for cash has driven villagers to increase their traditional slash-and-burn agriculture, as well as having forced them to attempt to grow cash crops such as coffee, cabbage, and exotic fruit for the export market. As a teacher I tried what I could to help them change the situation, and now, with the support of the project, I notice the villagers are succeeding in regaining some of their former self-sufficiency.

In the early days with FELS I began to appreciate the connection between these environmental problems and the social, economic and cultural situation of the villages. Having a good understanding of Karen language and village lifestyle, it was easy for me to sit with villagers long into the night. At first I seemed to do most of the talking, but after several months the villagers themselves started to take over. Through the project we arranged study or exposure trips to show Karen villagers other villages in the north and northeast where people were successfully practicing integrated and sustainable farming. It was rewarding to see the enthusiasm mount as new techniques were explained by fellow

Dealing With The Outside World: Women gather to discuss local concerns. They also learn the Thai language to deal effectively with encroaching threats to their land.

Photo: FELS.

farmers who had taken steps to deal with their own problems.

These trips fueled ever more lively and positive discussions once we returned. The first active step they agreed upon was to zone the forests in the villagers' area to help prevent further losses. Also, a few villagers were interested in trying alley cropping—planting leguminous trees along hedgerows, or alleys, with approximately four to five meters of food crops between each row—and contour farming. This would not only help to create better soil, but would also allow the same land to be planted every year. It took a few years but slowly other villagers began to see that using these techniques did indeed enrich the soil and reduce erosion.

After one exposure trip to see fish-rearing methods, three villagers decided to dig their own ponds. These were so successful that a year later there were thirty-eight ponds in the villages. Native fish are getting scarce in the nearby streams, so we hope their numbers will grow again now. Once ponds are dug, raising fish is fairly easy and extra fish can be sold for cash. Some women's groups are using extra fish to provide lunches for schoolchildren.

A few years ago a group of villagers expressed interest in planting fruit gardens, but they had no seedlings. The project helped by providing free seedlings (jackfruit, mango, tamarind). In some cases seedlings were not cared for. I discussed this with some villagers and we concluded that because the seedlings were free, they were perhaps taken for granted. The villagers decided to create a village fund for seeds. Seedlings were then sold to interested villagers at a reasonable price, and the fund money was used to extend fruit-tree growing to other villages. Now that some fruit gardens are established, villagers have decided to try growing native vegetables among the fruit trees. These permanent gardens have already helped reduce dependence on shifting cultivation.

We have found that helping villagers to identify their problems and the "why" of these problems is a process calling for plenty of time and patience. One of the first problems villagers mentioned was insufficient rice to last the year, forcing them to purchase rice or to look for alternative foods. A rice bank was suggested and quickly established—too quickly, we found out. Rice was freely borrowed but villagers were rarely in a position to repay. The bank lasted only a year.

We want to assist the people not only to understand these changes but also to encourage rediscovery of their own traditions, through sharing between youths and elders.

One of my tasks at the moment is to help facilitate the training of village volunteers, to allow them in turn to train other villagers in specific activities. Government health and agriculture officers are in the district, but seldom come up to the project villages. Having seen how well villager-to-villager sharing has worked, I am very interested to help further networks, not just between villages but between tribes, other nongovernmental organizations, and indigenous groups worldwide.

I am noticing many changes in the communities where I work. Young people are losing traditional beliefs. New values are evolving. We are sensitive to this and want to assist the people not only to understand these changes but also to encourage rediscovery of their own traditions, through sharing between youths and elders.

The villagers know this is their home and are making efforts to combat the problems they did not really create. I can only hope that government officials will see the villagers' efforts to protect the forest and develop a sustainable system of agriculture, and will understand these villagers are partners, not threats, in protecting the environment. We will be there to help present their case.

Part Two

FOLKS PLANTING SEEDS

While many ponder the consequences of global warming, perhaps the biggest single environmental catastrophe in human history is unfolding in the garden. While all are rightly concerned about the possibility of nuclear war, an equally devastating time bomb is ticking away in the fields of farmers all over the world. Loss of genetic diversity in agriculture—silent, rapid, inexorable—is leading us to a rendezvous with extinction—to the doorstep of hunger on a scale we refuse to imagine.

—Cary Fowler and Pat Mooney, *Shattering*

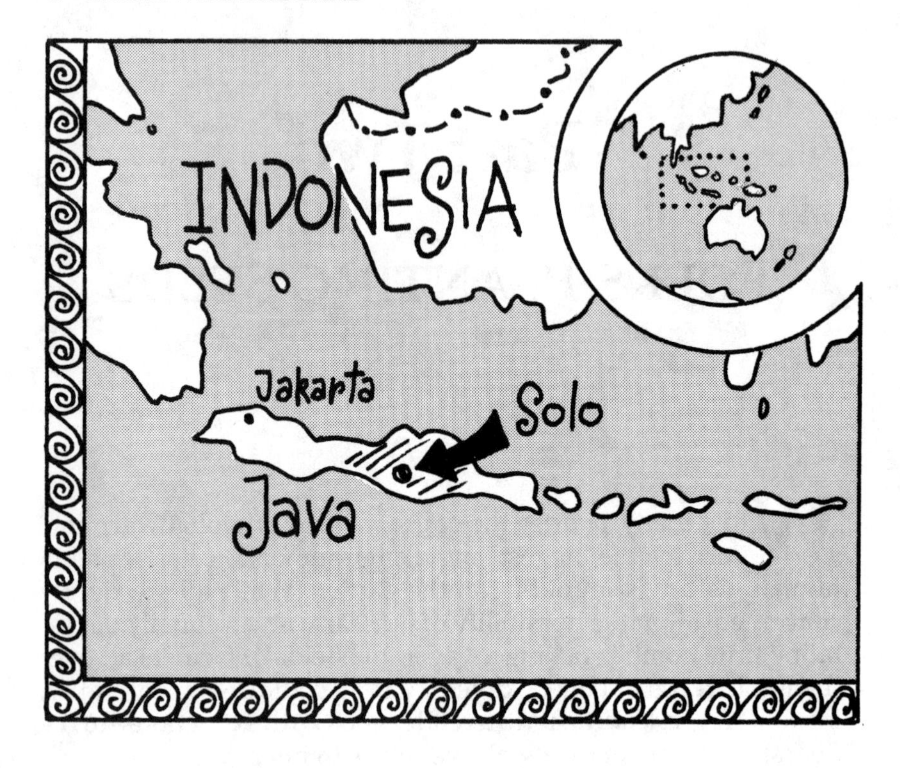

Contact:

Chandra Kirana

Ecological Studies Project
Jalan Kenanga 29
Badran SOLO 57142
Indonesia

tel and fax: (271) 45121

6

A Pocketful of Seed: Saving Old Varieties in Indonesia

Chandra Kirana

Reversing the effects of the "green" revolution has become a mission for many people involved in preserving seeds and culture. The creation of strains of high-yield variety (HYV) plants, intended to feed more people, has left us dependent on fewer and fewer plants for nourishment. Moreover, the new varieties only produce high yields when farmers use large amounts of expensive agricultural chemicals. The green revolution has also increased mechanization in agriculture, which tends to compact already-fragile soils, and cause a drain on scarce reserves of capital to buy farm machinery.

This "revolution" has not only made life less green in many areas, it has also helped to destroy a relationship between culture and nature, between diverse people and plants that have sustained generations.

Indonesia, one of the countries said to have benefited the most from the green revolution, is now self-sufficient in rice production. But Chandra Kirana of the Ecological Studies Project in Solo, Indonesia has a different story to tell. She is now involved in am attempt to regain some of the traditional foods and culture that have been lost through the distancing effects of modernization. For Chandra it begins with a pocketful of seed.

For centuries the Javanese have had an agrarian culture—so let me speak as a rural village child might. I am not a scholar. I speak from my heart. Sometimes it hurts to see so many books about my people and my culture lining distinctive libraries of the world, so rarely do these books speak in our voices—the voices of the Javanese people.

My people's culture is one whereby the old, by tradition, teach the young how to live with nature—it is a serene culture where balance and harmony are very important. For us, nature and culture encompass each other. If you see life as a whole entity, you cannot perceive only a certain section of it and ignore the other. Only by seeing the whole reality can one learn how to balance and create harmony.

The secret is that of never overdoing anything .... Never eat so much, for that will mean that many will have to go without in order for balance to be maintained.

The secret is that of never overdoing anything. Never soar so high, because that will mean many will have to remain far down below in order for harmony to rule. Never eat so much, for that will mean that many will have to go without in order for balance to be maintained.

You see, there is only a certain amount of food in this world. And we now produce this food without paying respect to the harmony of mother earth's tune, of her music, so that now she has almost forgotten how to sing.

The Seeds of Our Work

I grew up in a village and I know what it's like to be able to roam in gardens, by the river, and through the lanes to find food. Almost anywhere we looked, we could find leaves, beans, legumes, mushrooms, fish, and shrimp to cook for the day.

My work with women in the villages has shown me that so many of these natural food sources have disappeared. Village women today are buying much of their food, including vegetables. This shocks me and I want to do all I can to save what still remains.

The Ecological Studies Project (ESP) is the nongovernmental organization I work for. We are based in Solo, on the island of Java in Indonesia. The organization gave me the opportunity to write a proposal to HIVOS (Humanistisch Instituut voor Ontwikkelingssaminwerking), a Dutch funding agency which provided funds for the "conservation of genetic resources native to Java." Our project includes developing a seed bank in one village.

Goals and Strategies

The objectives of the project include the following:

1. To re-establish farmer control of seed at the grassroots level.

2. To preserve the diversity of food and medicinal plants, especially those native to Java.
3. To work toward a more sustainable mode of agriculture.
4. To provide a natural laboratory where we can study genetic conservation, propagation, and development.

In order to achieve these goals, we are pursuing several strategies, including:

1. Collecting seeds in central and east Java.
2. Propagating and developing those seeds on land to which ESP has access in the village of Sekaralas (on the border between east and central Java), where the program is now growing.
3. Organizing, raising awareness, and training with the community.

We are currently in the very initial stage of collecting seeds. The project is comanaged by an agriculturalist from ESP named Dewi, who is helped by a local farmer. I am in charge of making people in the villages more aware of the importance of preserving diversity.

As a mother with children, I know how women feel about getting food into their children's tummies. So I started by visiting individual women, always with an assortment of seed in my pocket. Our seed-bank project provided me with a valid reason to visit the village regularly and intensively. On these visits I could talk about anything with the women, but I would always bring their attention to how hard it seemed to be to find vegetables and other food plants these days. Many of these women are barely making ends meet, so they are very interested in this topic.

We lost so much of our diversity of hundreds of different grains and other living organisms, and mother earth lost much of her fertility. But what we are realizing more and more is that we have lost a large part of our self-identity, our culture.

After a few visits I determined who were the most approachable women and started to frequent their homes more often. Other women would come to chat, too. In discussions about food cooking, the health of the children, and money difficulties—all good topics to talk about if you want to get close to women in the village—I would mention how different it was when *we* were children and were able to collect the day's vegetables and proteins from the land around us. We all started to pay more attention and as a group we began to realize how many varieties of food we used to eat as children were no longer available. We began to try to reason why.

The Roots of the Problem

In the early 1970s, our villages in Java began to encounter the silent bewilderment of the green revolution. We began to produce high-yielding varieties of rice, which could withstand strong winds because they were so short. Strange that we never stopped to think that we rarely had strong winds anyway. We were introduced to chemical fertilizer that worked wonders, and to wizardly pesticide sprays. Now we could plant rice three times a year; each crop would last a hundred days. By 1984 Indonesia, our country, had become self-sufficient in rice production. A wonder! That is true, but at what price and for whom? True, some of the poor of the world were fed, but to ease whose guilt? These are but a few of our problems.

We lost so much of our diversity of hundreds of different grains and other living organisms, and mother earth lost much of her fertility. But what we are realizing more and more is that we have lost a large part of our self-identity, our culture. My generation, the children of the 1960s have forgotten how to fly to the music of nature. Some of us have learned to dance to the bewitching tones of modernization, it's true. But the majority of us are lost like birds with broken wings—a process in which the green revolution was an important factor. This is especially so for the village children of Java.

The Losses

Slowly the green revolution has displaced women and children more and more from the agricultural production process. Industrialization has helped farmers to become more mechanized. Many people in the village thought they were doing their children a favor by not exposing them anymore to the work and heat of the fields, and so the children never learned. Children were sent to school. I remember how we were taught to aspire to become "clever people," like the people in the cities. We were innocently unaware of the fact that our parents did not have the money to send us all the way, so most village children went as far as junior secondary school and remained ignorant about working the fields. We have not learned how to read the stars and the signs that nature generously gives us as guidance. We have even forgotten how to listen to the cries of our own bodies.

There are many other losses. In the past you could practically eat a village hedge, as it usually consisted of different edible plants. Now, due to compulsory village neatness competitions, you find most fences are made from dead bamboo and nails. Fish in the river and mushrooms are mostly dead because of the poisons used so extensively now. We notice that the delicious wild spinach which used to grow in abundance and was so very nutritious, has now become so rare that you cannot go out with a basket and collect some for a meal. This spinach in Javanese is called *bayam kandang*, meaning "stable spinach," because it grows near stables where there is a lot of dung. Before chemical fertilizers were available in the village, people used dung as a fertilizer and so the seed of this spinach went everywhere with the dung. It would grow abundantly and, as it didn't hurt the main crops, farmers never weeded it out. Now dung has been replaced by a chemical fertilizer that kills the bayam kandang too.

Another nutritious food source, the winged bean (*kecipir*) has also become extinct in this village. One woman recalled that she still had some a few years ago, but because the children really liked the beans, the family ate all they had and didn't think of leaving some to grow old for seed. Now she was sorry and wondered if she could get hold of some seed, if the plant still remained in another place. I had some *kecipir* seed in my pocket and was happy to give her some, providing that she let it grow into seed for herself and anyone else interested. We decided that we are responsible for providing seed for ourselves and our children.

Deschooling and Rediscovery

To date, only three women of the village are planting seeds they have requested, but neighbors are watching and joining in monitoring the results. When I cannot provide the seed, I try to encourage women to look for them in the area, to see if maybe some have survived. Many are already eager to do this, as they are seeing the benefits.

We are deschooling ourselves and trying to learn, together with the community, how to be more sustainable and self-reliant.

Time, dedication, and love are all essential ingredients for growing seeds. Water is also very important. Sadly, at the moment we are facing a very harsh drought, and only those who live close to a well are able to grow the seed.

A more sophisticated seed bank is planned for the overall community program. Ideally, community members will be able to go there to learn how to make compost, how to propagate seed or trees, how to deal with pests, how to do companion planting, and so on. Dewi and I and all the other ESP people are also still learning about these things. At this stage we are learning skills to develop more sustainable agriculture. We have just worked out how to make compost, as well as a few pest-management and companion-planting techniques.

Although many ESP people were trained as agriculturalists, they learned from a different perspective—one that focused on how to use chemicals. We are deschooling ourselves and trying to learn, together with the community, how to be more sustainable and self-reliant.

Banking on Seeds

We are trying to model the seed-bank and farm program on the traditional Javanese home garden or *Pekarangan,* where one finds tall coconut and fruit trees, edible creepers, tubers, herbs, medicinal plants, vegetables, animals such as chickens and goats, and a fish pond. Based on this model, every farmer and gardener will also be a guardian and curator of seeds. Through the network of villages where ESP works, we will be able to make seed available in a larger circle and maybe replicate the program appropriately in other areas.

Now we are still at an early stage, having just had our first birthday. But as concerned individuals contributing our own experience and knowledge, we have been around for quite some time. Here's hoping that joining forces as a group within ESP will bring about good results. In time, we may relearn the wisdom of our ancestors—in connection with nature.

Giving Back

Chandra Kirana

We talk of giving life back to our mother earth
But a friend of mine
envisaged an old, frail and toothless woman
Who was so sick, she could receive no more ...
I thought it was so true
Then I see our children ... so tender ... so young ... but so strong and full of hope,
I know the only hope lies in our ability
to teach them how to give
Give from the heart, with love from soul
Only then will mother earth be saved
For this our generation will never be able to do
We ... have never learned how to be generous
Society has taught us only
Of how to become maestros in the
Art of unquenchable wanting.

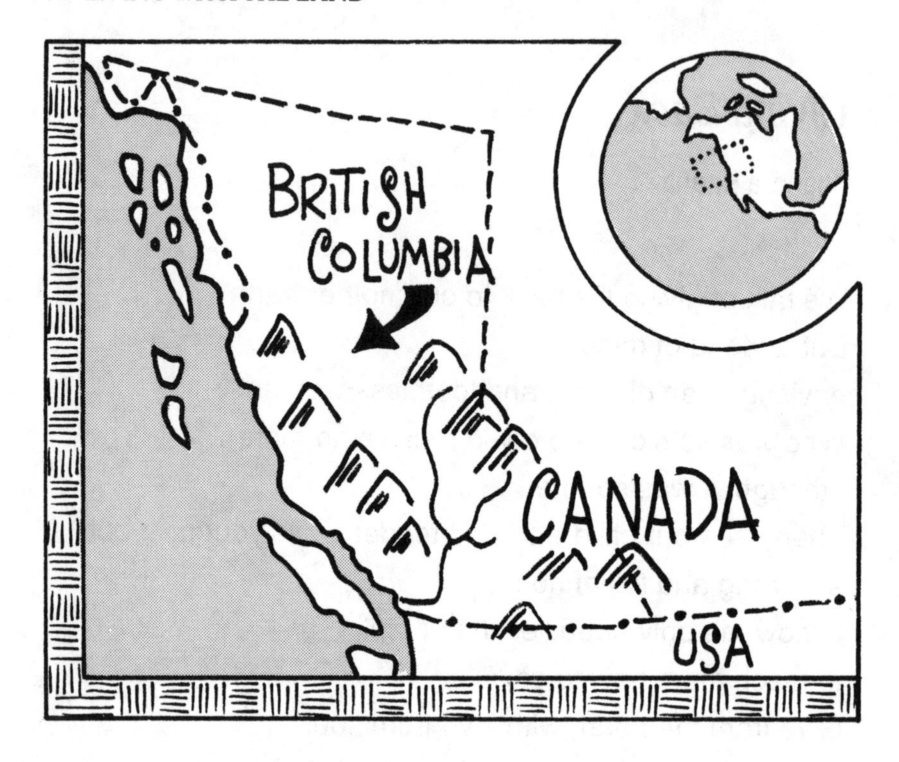

Contact:

Thomas Evans

Planting Seeds Project
c/o Circling Dawn Organic Foods Collective
1045 Commercial Drive
Vancouver, British Columbia
Canada V5L 3X1

tel: (604) 255-2326

7

Putting Culture Back into Agriculture: Collective Action to Save Seeds

Thomas Evans

In the last century farming has evolved from a small-scale, labor intensive, family-based activity to the mechanized mass production of food known as agribusiness. It is no coincidence that in this same period, the world has lost thousands of varieties of plant foods. Companies involved in agribusiness have extended their operations into all stages of food production, including the alteration of the genetic structure of seeds. In doing so, agribusiness has marginalized not only farmers but also the culture and practice of farming, which includes saving and exchanging seed from season to season and from farmer to farmer.

The modern farmer is forced to buy new seed each year because companies are creating plants that cannot reproduce themselves reliably from their own seed. These genetically manipulated plants also require chemicals to grow successfully. The farmer becomes entangled in a chemical-dependent, mechanized system that has resulted in the unprecedented depletion of precious topsoil, water contamination, erosion of the gene pool of human food crops, and corporate control of plant life itself.

Thomas Evans is a founding member of Circling Dawn Organic Foods Collective in Vancouver, British Columbia, a group of people who live together and maintain the only completely organic foodstore and restaurant in Canada. With support from the collective, Thomas created the Planting Seeds Project in the spring of 1991. He was able to convince John Kimmey, founder of Talavaya Seeds in New Mexico, to invest some of his time in the project. They traveled together to rural communities in British Columbia to raise awareness about the diminishing gene pool of our food crops. This work involves planting practice gardens of open-pollinated (as opposed to genetically engineered) seeds, while engendering a relationship of mutual stewardship between seeds and people.

Where did my involvement with food and seeds begin? Really it's been a process of awakening to a deeper understanding of why I'm alive. I've learned that my relationship to my ancestry is also a relationship to what I eat. So when we talk about seeds, we're talking about how we've fed ourselves through time and we're also talking about culture. That's where the old word *agriculture* came from. Seeds had to be cared for, harvested, and saved for the next year. Up until about fifty years ago, the fact that we had any food to eat at all was totally dependent on the unbroken chain of events that occurs in culture, that takes care of these seeds and passes them on from generation to generation.

That process is all changing. Now we are under the pressure of multinational agribusiness systems. Petrochemical and pharmaceutical companies involved in hybrid seed production are acquiring rights and patents so that they can control the genetic pool of various plants. Once hybridization and patenting has occurred, they destroy the parent plant and then they have complete ownership of a form of plant life. That is how 70 percent of the original genetic pool has been lost in the last fifty years. How long will it take for the last 30 percent?

People are becoming aware of the effects of chemical fertilizers and pesticides and are realizing the importance of food grown without chemical inputs. By paying more for organic food, they are doing an important environmental service, but I don't believe they are getting any more nutrition, because this comes from the genetic structure of the plant. Seed companies are tampering with the genes of plants, altering their design to improve the size, color, taste, and look of uniformity; and, I believe, plants are losing nutritional quality in the process. Seed companies see the trend to buy organic food and think it's a fad. They sit back and say, "Let them do their thing;" they know that attempts to grow food without chemicals are all for naught because we don't control the seeds. And they can program the genetic structure of plants to require chemicals at any time.

How It Began and What Keeps Us Going

In the process of creating the Planting Seeds Project, I got in touch with a man named John Kimmey, who had worked with indigenous people and on saving open-pollinated seed in the United States. I intended to take some traditional seed up to the Lil'wat and Shuswap territories,hoping to put native growers and activists in touch with some of the Canadian farmers I worked with through the store. John said he could be of help. We combined resource materials, loaded up his car, and headed out with enough gas to get to our first destination.

One of the stops on our first trip was in traditional Doukhabour country, in the Kootenay region of British Columbia. These people have a strong culture that, like native culture generally, has been gradually damaged by assimilation. We established an immediate link with the younger people from the community, who, because of the way they were brought up, seemed to know exactly what we were talking about. The older folks were very skeptical, and it took some doing to get them to come and talk with us or to let us visit them. But once we had connected and they saw what we were doing, magic started to happen.

Eventually we were told about Walter, a 71-year-old man who had been practicing traditional farming all of his life. On Sunday we met each other. Walter had an amazing set-up on his farm. He only bought three commodities from the store—oil, salt, and coffee. Everything else he grew himself. This was around June 1, and he was eating his own tomatoes and romano beans and his peppers were just about ready. He had his own glass greenhouse and he had built bunkers out of cement for growing beds. Although he lived high up in the mountains, he had to harvest early because he would dig out the bunkers in January and then build a compost in the bottom two-thirds. He'd then put topsoil in the top third and the compost would start to cook. From January to March the compost would heat his greenhouse and keep the soil warm enough for the plants to grow. Walter was growing mostly open-pollinated seeds, which had adapted to his growing style over the years. By the time the compost had cooled down and the roots had reached into the nutritious soil, the sun was high enough so that it had taken over heating the greenhouse. Walter was eating fresh vegetables all year round.

Suddenly, the funny old guy down the road was seen as the resident expert in the field that the new generation of families were becoming involved in. The young folks we are working with now can no longer avoid understanding people like Walter. A scene that is etched in my mind took place while we drove away from the community. I remember looking into the rear-view mirror and seeing Walter with his big white beard and Neta, a young woman who had hosted our workshop, walking in her garden and pointing at this and that and talking. It is amazing to think that we might have had a hand in the process of bringing the two together in a working relationship. This kind of experience sustained us on the whole trip.

A Typical Workshop

I would begin workshops with a prayer that acknowledged the indigenous life and spirituality of the area. John would then give a warning about the spirit and intention that lives within the seeds, and how this should not be tampered with by human beings. Then I'd go into a history of the evolution of the earth and humanity from an agricultural perspective. I'd also discuss the native-sovereignty movement and our responsibility as Canadians to understand native elder culture and respect for territory. After that, we'd show John's video on Hopi prophecy and end with a talk about the organic movement and the importance of whole plants and whole seeds.

If someone asks how much it costs to start doing this, a truthful reply is that it costs almost nothing except the rest of your life.

The next part would begin with a discussion of how open- pollinated seeds work—developing the idea that they are alive and that you can communicate with them. John would present a slide show of the farming techniques he used in his seed-banking work in New Mexico. After that, we'd get to the hands-on work, where we'd plant thirty or forty varieties of seed into a 20-by-20 plot. The person tending this garden will spend the year observing the plants, keeping field notes, and trying different experiments like singing to some, shading others, spacing, or mulching. It's the idea that you learn from anything by waking up to the fact that it is alive.

Our Goals

We put in eighteen gardens on this trip and we only covered one section of the province. At harvest time we will visit these people and their gardens again, as well as going to some new areas. We'll help with the harvest, review the notes, collect the seeds, see what grew well, and make plans for what to try next year.

The second phase involves going into higher production for more seed and starting on a new level of experimentation, maybe specializing and focusing in on varieties that really worked well. We'll run half a dozen different cover crops in a structured way over that 20-by-20 plot and start learning and refining techniques in order to maximize production and efficiency—all the things that are important to commercial

Diversity On Trial: Members of the Circling Dawn Collective, and farmers from the Stein mountain region in British Columbia, have planted a diversity trial garden.

Photo: Circling Dawn Collective.

farming. We'll collect field notes for the second year with those plants, while trying another thirty or forty different varieties to expand our understanding of what works and doesn't work for a particular area. By the third year, the few seeds that really worked well will be isolated and distinguished. By then, we hope the relationship between the Planting Seeds Project, the grower, and the plants will be concrete, and the farmer will be able to grow a significant portion of open-pollinated seed.

While that seed is being collected, we will be working to get the other end of the system—the marketing strategy—in place, finding seed companies that are ready for bulk handling of open- pollinated seed and farmers ready to work with them. We want to get to a point where farmers can say, "OK, I'll grow open- pollinated squash this year instead of hybrid, and I can do it because the seed has been worked out for my area and there is enough of it and it's been proven workable and I can get extensive field notes and information from the seed grower."

From Soil to Market

The farmer needs a guarantee that the open-pollinated seeds and produce will sell as well as the hybrids they used to grow. As with any transition, it is important that the current practice can stay in place until the new one being developed is strong enough to take over. It's hard not to panic. We're dealing with a catch-22 situation, in that we're not going to have open- pollinated food or seed until there's a market for it, and we're not going to have a market for it until there's a seed to begin with. So what comes first? It's a complex, creative trick to get them both happening at the same time.

Seeds cost very little. You take a handful of seeds and give it away to someone, and if that person with integrity and sincerity grows out that seed, they'll have hundreds of handfuls.

If someone asks how much it costs to start doing this, a truthful reply is that it costs almost nothing except the rest of your life. Because once you embark and you have that little packet of open-pollinated seeds, they are relying on you. You have a relationship, you've made an agreement with them and you want to be careful of that. There are a lot of resources to help you get started in seed saving, and it doesn't take a lot of room to practice. You don't have an excuse like, "I don't live on a farm." You can even do it in a flowerpot, because it's all about practicing and at least you're getting to know the plant as you watch it grow and flower.

What is more, seeds cost very little. You take a handful of seeds and give it away to someone, and if that person with integrity and sincerity grows out that seed, they'll have hundreds of handfuls. One handful can turn into big sacks just like that. It can turn into enough seeds to start self-sufficient gardens of that particular variety in many places. But it's a bit of a race. It's got to spread itself out to the four corners of the earth faster than big companies can destroy it.

8

The Plant Breeders' Rights Act: What it Really Means

GROW (Genetic Resources for Our World)

The passage of the Plant Breeders' Rights Act in Canada in June of 1990 means that this country will allow the patenting of the genetic material of higher life forms, in other words the blueprints of life itself. Until recently, genetic material was universally considered a public resource and not private property. It was believed that seeds and other reproductive material, even if altered through selective breeding, are products of nature. They were considered sufficiently different from machines and mousetraps to be outside the domain of patentable products. Companies sold seeds, but they did not own the plant varieties they sold, and they had no exclusive rights to market them.

During the first quarter of the twentieth century, things began to change. Plant breeders began to view their new varieties as inventions and demanded that they be allowed to benefit from their work by collecting royalties. In 1930 the United States passed the Plant Patent Act, and in 1934 the first of a series of plant patents was granted in Germany.

Canada has now joined the relatively small number of industrialized countries that recognize plant patents. The central thrust of the Plant Breeders' Rights Act is to create "rights" that are virtual patents—giving plant breeders the exclusive rights to market the plant varieties they develop.

The Wrong Model of Development

The new science of genetic engineering or biotechnology is one of the major scientific developments of our time, and few would deny that it holds great promise for agriculture. There is a dire need to increase the productivity of farmland around the world in order to feed a global population expected to reach ten billion early in the next century. Biotechnology offers the hope of being able to do that by creating more nutritious crops, crops resistant to pests and diseases, and varieties that are suited to adverse growing conditions. The question is not whether to proceed with biotechnological research, but how this research is to be conducted, for what purposes, and by whom. In other words, who is to control the research and decide what should be developed and introduced into the environment?

The GROW coalition maintains that in passing the Plant Breeders' Rights Act and introducing seed patenting, Canada is opting for precisely the wrong approach. It is relinquishing society's control over this powerful new technology and placing it in the hands of people whose primary concern is making money, rather than addressing the increasingly urgent tasks of environmental protection, the eradication of world hunger, and the preservation of small-scale farming, upon which so many people depend for their livelihood.

Control in the Wrong Hands

The new law inevitably places much of the control over biotechnology, and thus the future of world agriculture, in the hands of multinational seed companies—many of which are in turn owned by chemical companies.

In 1988 the *Economist* reported that chemical companies "have spent $10 billion or so in as many years buying up seed companies world wide." The world's largest seed companies are Pioneer Hi-Bred, Sandoz, Dekalb-Pfizer, Upjohn, Limagrain, Shell Oil, ICI, Ciba-Geigy, Orson, and Cargill. Seven of these ten companies are also involved in selling fungicides, herbicides, insecticides, and chemical fertilizers.

An Environmental Time-Bomb

It is naive to suppose that these companies will develop seed varieties that will promote environmentally sound agriculture. They are in the business of selling agricultural chemicals, and it is in their best financial interests to promote high-tech, capital-intensive (read *expensive*) agricultural practices. More than 50 percent of all seed research conducted by private enterprise at the present time is devoted to developing crop varieties that are tolerant of high concentrations of fertilizers, pesticides, and so on, rather than varieties that are themselves hardy and pest resistant.

At a time when more and more concerns are being raised about the presence of toxic chemicals in the food chain, the Plant Breeders' Rights Act emerges as a kind of environmental time-bomb. The bill encourages the escalating use of chemicals in agriculture, with as-yet-unknown effects on human health. Intensive, high-input farming, using many chemicals, is increasingly being linked to widespread land degradation and ecological catastrophe.

Victimizing the Third World

The crops developed by multinational seed companies are more suited to the industrialized nations, where there are large tracts of land with comparatively few people to work them. In most developing countries there is less land available and larger populations that have traditionally made their living by farming, using labor-intensive, hands-on farming practices. Corporate farming, increasing reliance on a single crop (monoculture) and the increased costs of high-tech agriculture—all encouraged by the concentration of power over agriculture in the hands of chemical and other multinational companies—are literally driving people off the land. They are migrating to already-overcrowded cities, where they often can find neither work nor shelter.

Third World countries are having to pay royalties on seeds that were developed in the industrialized countries of the north, from genetic resources that originated in the developing countries of the south. Virtually all of today's major food crops were developed from genetic resources freely donated by developing countries.

As author and agricultural economist Pat Mooney has said, "It doesn't thrill Third World countries a whole lot to think that their genetic diversity they've given to us is now subject to patent." The Plant Breeders' Rights Act is an extension of the classic North-South economic relationship, which sees raw materials flow cheaply to the North, only to be sold back to the South as expensive, "manufactured" goods.

Threat of Food Scarcity

Because of the increasing dominance of multinational seed companies in agriculture, the world is becoming dependent on fewer and fewer crop varieties. The danger inherent in this is that at any given time a single crop can be wiped out by disease or by a pest that has developed resistance to the pesticides currently in use. Increasing dependence on fewer crop varieties exposes us to greater susceptibility to widespread crop failure and, consequently, famine. In his book *Altered Harvest*, author Jack Doyle makes the case that corporate control over agriculture, concentrated in ever-fewer hands, courts global disaster.

Such a scenario is all the more plausible because high-tech farming requires a constant development of new crop varieties. Yet, at the same time that we are encouraging high- tech farming around the world, we are losing the "gene banks" on which future varieties depend. The bulk of the world's genetic resources are located in the threatened tropical ecosystems of Third World countries—such as the rainforests currently being razed for development by debt-ridden countries. It is estimated that one species of plant or animal is lost to humanity every day. Future biotechnologists may not have the raw material they need to develop the new crop varieties necessary to feed humanity, after pests and diseases have adapted themselves to all of our new, "superior" crop varieties and agricultural chemicals.

The Brundtland Commission on Environment and Development states that "research and development, production, and marketing need to be carefully guided so as not to make the world even more dependent on a few crop varieties—or on the product of a few multinationals."

For more information on the Plant Breeders' Rights Act, contact the Rural Advancement Fund International (RAFI), or Genetic Resources for Our World (GROW), both at: Suite 750, 130 Slater Street, Ottawa, Ontario, Canada K1P 6E2; telephone (613) 565-0900, or fax (613) 594-8705.

Part Three

FOLKS IN THE FIELDS

It is better to sow an idea in the heads of a hundred, than a hundred ideas in the head of one.

—Eric Holt, advisor to UNAG

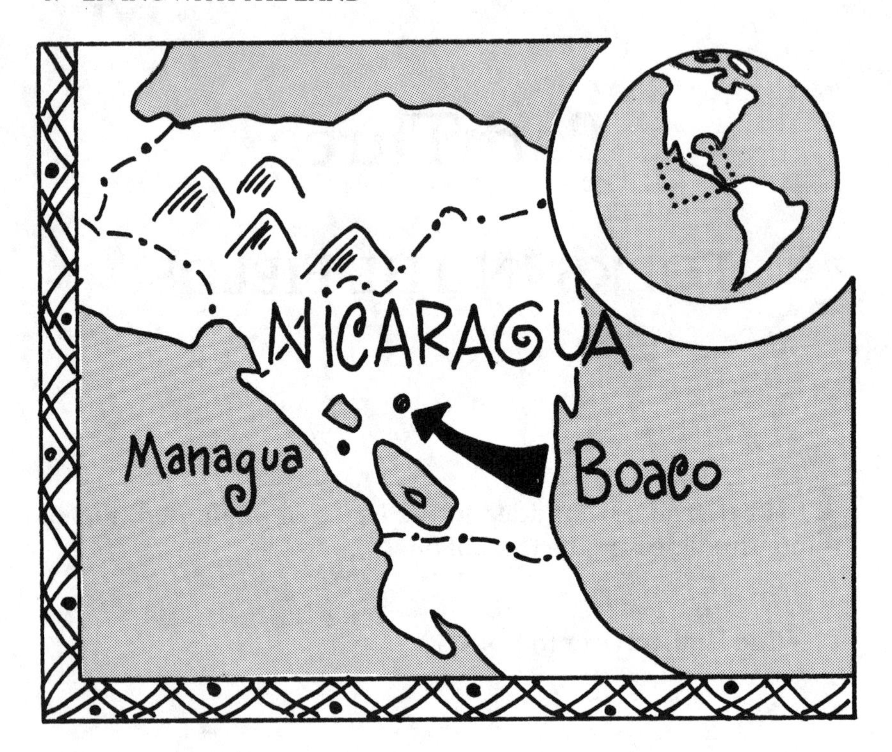

Contacts:

Robert Fox

Oxfam Canada Representative for Nicaragua
Apartado 4929
Managua, Nicaragua

tel and fax: (5052) 40796

Marcial López, Jorge Iran Vasquez

Programa Campesino a Campesino
UNAG
De Plaza Espana 2 cuadras abajo
Frente a las oficinas de Naciones Unidas
Managua, Nicaragua

tel and fax: (5052) 661240 or 664974

9

Learning to Love the Land Again: Campesinos Empowering Campesinos

Richard McDermott

At the heart of many communities is the need for people to regain control of their land and their dignity. In Nicaragua, agrarian reform has given many rural people legal title to their land, but their struggle now is how to regain the soil. This story is about the attempts of campesinos (farmers) in the dry zone to reduce soil erosion on the steep slopes where they farm. In the words of one farmer, "we do this so that our children may inherit soil rather than stones."

This story comes from interviews conducted by Richard McDermott, who went to Nicaragua to see the program in action. Richard met with Marcial López, who works with the National Association of Farmers and Ranchers (UNAG) as part of the national team that promotes the Campesino a Campesino program. He also went to the dry zone in Santa Lucia, where the project has been quite successful. There he talked with, among others, Enrique Mendoza and José Jesús Mendoza, who are rural promoters. These interviews offer valuable insight into the grassroots cooperative movement and reassert the importance of this program in decreasing the Nicaraguan farmer's dependence on damaging agricultural practices.

Marcial López: Managua

Our work is in practical teaching and doing. We go to the farmers and help build up their knowledge and skills. For example, they learn how to conserve soil through the use of organic fertilizers.

Although the national team is based here in the city, we are active in all parts of the country. We are active in Estelí, in Huigalpa, in Rio San

Juan, in Rivas, in Matalgalpa, in Santa Rosa del Peñon, and in Santa Lucía. In each of these places there is a team of campesino promoters—we have two hundred promoters in all. Together, we conduct workshops and field trips. On a field trip we might go together to a particular plot of land to analyze the specific problems or challenges there and how they might be resolved. The key to this program is the sharing of information. The promoter is the person next door; you see them make mistakes and also do things well.

As members of the national team, we work from five to seven days a week. We usually meet with the participating campesinos on the weekends, since that is when they can spare time from their work in the fields. We visit each group of participants perhaps twice a month—there are many to visit!

The key to this program is the sharing of information. The promoter is the person next door; you see them make mistakes and also do things well.

There are about two thousand campesinos who use the methods that we are developing. In spreading the ideas and techniques of the program we have to be very flexible, because Nicaragua is a diverse country. There are no rules, just examples. Each campesino has to rely on the resources they find themselves with. In Santa Lucía, for example, the soil is full of rocks, so these can be used to build barriers to prevent erosion. However, in the east rocks are harder to find, so barriers must be made of a row of some sturdy crop with deep roots.

Among those we work with are ten groups of women who have organized themselves into collectives. They grow a whole range of vegetables such as carrots, cabbage, and beans, using methods that conserve the earth, like composting, terracing, and intercropping varieties of plants.

Our most advanced project is in Santa Lucía in the municipality of Boaco. It's been going for four years, and in that time the population has managed to reduce the use of agro-chemicals through reliance on organic methods, and at the same time increase their production. With the use of manure, compost, and mulch, they've conserved both soil and water. Because of the clear success of these techniques, lots of people are adopting them.

In Santa Lucía, the group of farmers who first participated in our initiative are now organizing a cooperative. They are in the process of negotiating to buy themselves silos, which would be an important step toward economic sustainability. When they have silos they will be able to store the harvest, to sell it when the time is best.

Also in Santa Lucía, they have formed a campesino school. There they offer instruction not only to people from the area but also to campesinos from the rest of Nicaragua. Those who teach are the campesinos themselves. Our biggest challenge in Santa Lucía has been to meet the high demand for instruction. We don't have sufficient facilities to teach all those who want to learn the methods.

The Campesinos and Promoters: Santa Lucía

Every year we used to burn our land and use chemical fertilizers and insecticides. These practices destroyed all the creatures that added richness to the soil. But not all the creatures are our enemies. Why destroy them all, when most are good for the soil? The result is the same as the war that Nicaragua suffered for so many years: among the guilty, many innocent die as well. Besides, when you burn their food supply, those that survive have nothing to eat except the single crop you've planted. You are forced to apply insecticides to keep them under control and prevent plagues from developing.

There are three reasons why we no longer use chemicals. They impoverish the soil by reducing the diversity of creatures within it, they're too expensive, and the harvest that results is of less nutritional quality. The people much prefer the food that we now grow organically. Chemically grown food doesn't have the same vitamins as our organically grown food.

Now when we see other farmers burning their land, it hurts us inside. How ungrateful! How crazy to become dependent on fertilizers by destroying all the organic material that adds life to the soil! Now we are pleased to see creatures in the soil, because we know that they aerate the soil and help decompose the organic material we plough under the soil rather than burn.

In Santa Lucía there's an additional problem. It is in mountainous country. We used to plow in straight lines without consideration for the contour of the land. Gradually we noticed our soil was vanishing. Rocks began to grow out of the soil better than our crops. After heavy rains we could see our soil running down the main street of Santa Lucía. We were losing our soil to erosion and our harvests were becoming smaller year by year.

And then we met with the Mexican promoters. They taught us to respect the land again. Some of us had been working the land for thirty years without knowing the soil, without understanding what we were

The Curve Of The Land: Promoter, Augustin Bello, teaches the others about finding the contours of the land using the A-frame method.
Photo: UNAG.

doing. We learned that we had to nurture the land, not just exploit it. We stopped burning refuse and instead we ploughed it under. We started using organic fertilizers that we created in our compost heaps. We make organic fertilizer from waste material, manure, water, and soil. This fertilizer has all fourteen essential elements that our crops need, rather than just the three (nitrogen, potassium and phosphorus) that chemical fertilizers contain.

Plants are living beings as well. If we as people are feeling sick, we must nourish ourselves or we couldn't survive. The same is true of our crops. They also need to be nourished.

We have discovered a new plant, the "velvet bean," which can be grown alongside the crop we've planted. It eliminates the need for weeding, prevents erosion, reduces evaporation by providing ground cover, fixes nitrogen in the soil, and produces a huge quantity of refuse, which adds nutrients to the soil.

The Campesion a Campesino Program

Greg Utzig

Prior to the Sandinista triumph in 1979, the dry zone of Nicaragua had always been an economically poor area. The area has suffered severe environmental degradation due to overcutting of trees for firewood, overgrazing by cattle, and the attempts of marginalized people to farm the steep, vertical slopes of the volcanic hills. Initially the Sandinista's plan was to relocate people to more fertile areas in the north. However, during the contra war the people were attacked and so they began to migrate back to the area. It was concluded by UNAG, the National Union of Farmers and Ranchers, that it might be better to look for improvements in farming techniques that would allow people to live in the dry zone. This was the seed of the Campesino a Campesino program, which has its basis in the idea that peasants learn more effectively from other peasants than from technicians or professionals.

The program originated in Guatemala, where agronomists worked with indigenous people to improve soil fertility in their mountain plots. The success of this made the program and its participants the focus of attack by threatened land-owners. However, the efforts of these people did not go unnoticed. These same techniques have spread to other Central American countries. The project in Mexico provided six weeks of training for a team of Nicaraguans. From there they returned to their communities to teach others what they had learned.

The Campesino a Campesino program has two aims—to increase the productive capacity of the soil while reducing dependency on herbicides, pesticides, and fertilizers. In Mexico the team learned ways to ensure that these aims are met. This included training in contour farming, agro-forestry, building composts, and mulching.

In Nicaragua the program has met with tremendous success, for the very reason the Guatemalan one didn't—land tenure. The feeling of security that comes from knowing they have the right to their land, allows campesinos to invest more energy into innovating approaches to agriculture. Robert Fox, a development worker with Oxfam Canada in Nicaragua, says that even though the current Nicaraguan government has begun a legislative attack on agrarian reform, the Campesino a Campesino program has not been significantly affected. To the contrary, it has grown incrementally over the last year, attempting to provide a foundation where other rural supports have been withdrawn.

The Mexican promoters also taught us how to prevent erosion by building ditches and terraces. Rocks used to be a scourge, now they are cherished. On steeper slopes we build the ditches closer together, and on the steepest slopes, which are the most susceptible to erosion, we just build terraces and no longer work the land intensively, but plant fruit trees instead. We now plow the land in such a way that if you were to walk along the furrow, you would neither rise nor fall.

This conservation work requires a lot of effort. Everything is done by hand. But now when it rains, our ditches, terraces, and furrows retain the water, which is allowed to seep into the soil, and we no longer lose our soil to erosion.

We now understand our land better and respect it once again. Plants are living beings as well. If we as people are feeling sick, we must nourish ourselves or we couldn't survive. The same is true of our crops. They also need to be nourished.

Now we feel so proud to be campesinos. We recognize that much depends on us. In Nicaragua, small-scale farmers such as ourselves feed the cities.

The result of all our efforts has been two- or three-fold increases in yields at harvest time and the knowledge that our land will continue to be productive into the future. We are now self-sufficient. We no longer need to go to the bank to beg for money, for since we stopped using chemicals we have reduced our expenses considerably.

We have a new sense of pride. Before, when people saw us as campesinos, we didn't feel good. Now we feel so proud to be campesinos. We recognize that much depends on us. In Nicaragua, small-scale farmers such as ourselves feed the cities.

10

On the Edge of a Disappearing Forest: Traditional Farming Systems in Transition

Lucy and Larry Fisher of World Neighbors,
Stefan Wodicka of CUSO Indonesia,
and Putra Suardika of Yayasan Tananua

A farming system that has sustained rural upland communities for generations is slowly eroding. Farmers in many countries are finding it increasingly difficult to find new land to cultivate while waiting for their old fields to regain fertility. The practice of shifting cultivation is being replaced by more permanent farming methods, through techniques developed and adapted by farmers assisted by development groups. In this way, farmers are continuing to meet their basic needs while protecting the surrounding forests. The techniques being adapted by Yayasan Tananua, an Indonesian farmer organization, combine traditional and modern ideas that are passed from community to community through a growing network of concerned farmers. Here is how it works.

The island of Sumba, along with the rest of East Nusa Tenggara Province in Indonesia, lies in the rain shadow of the Australian continent. Here, the rainy season is unpredictable and the soil often poor.

Farmers in the rugged uplands of eastern Sumba are generally shifting cultivators, planting maize, sorghum, cassava, upland rice, and beans, or peanuts on hillsides or in valleys. Farmers open land by clearing and burning, after which crops are planted in the ash. After two

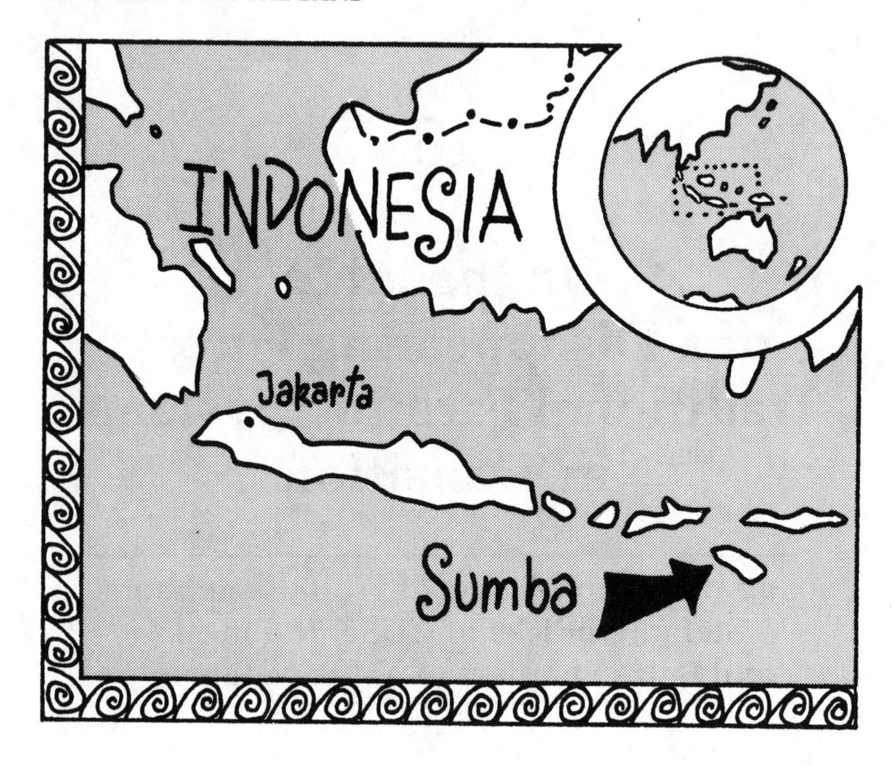

Contact:

Patris da Gomez

Yayasan Tananua
P.O. Box 1107
Kupang, NTT
Indonesia

Larry Fisher, South East Asia Representative

World Neighbors
P.O. Box 71
UBUD 80571
Bali, Indonesia

tel: (62) 361-95273
fax: (62) 361-95120

to five years, however, the land becomes eroded and infertile, and yields drop; the land must be fallow for ten or more years before fertility returns.

Unfortunately, increased population pressure is making this system of agriculture unsustainable. Today, the forests of Sumba are largely gone: the clearing of land on steep slopes has resulted in soil erosion and in grassland overtaking forested areas.

But the people still depend on this now-degraded and elusive forest. It has traditionally provided fuelwood, fodder, construction materials, medicinal herbs, spices, oils, dyes, and food. When there is drought and crop failure, the forest becomes especially important. During famine and times of scarcity before the harvest, the forest provides a major source of livelihood, including the famine tubers, an important last-resort food that must be cut into pieces and then rinsed for several days in the river to wash out the toxins before being consumed.

The Sumbanese farmers are becoming aware that the era of shifting fields and forest harvesting is drawing to a close. But it will be difficult to change a history of dependence on the forest if alternative sources of food and wood are not available.

Faced by poverty and the destruction of their environment, Sumbanese farmers in several villages sat down together in 1981 to talk about alternative solutions. The discussions were facilitated by World Neighbors, an international development organization. Farmers described their major problems as low yields, limited incomes, and changing land-tenure rights.

Several years later, leaders from the resulting dry-land farming program formed an autonomous organization, Yayasan Tananua. The name *Tananua* is taken from the lines of a Sumbanese proverb, *Tana nua, watu lihi*, which means "land that shares a common boundary, rocks that separate and bind." Today, nearly seven hundred farm families in thirteen villages in the uplands of eastern Sumba work with Yayasan Tananua.

From Shifting to Sedentary Agriculture

The farmers realized they needed first to address the problems of soil erosion and loss of fertility. A few farmers agreed to try out some basic soil and water conservation measures in order to sustain or improve crop yields. The technologies developed and adapted in recent years by Yayasan Tananua are based on traditional knowledge. For example, Sumbanese farmers have long known the strategy of contour farming and about the use of the *kawia* tree (*Albizia Chinensis*) as an indicator of soil fertility.

Pointing Toward The Future: Looking at practical guides to dryland farming together.

Photo: Larry Fisher.

Building upon this local knowledge, a system was developed for planting crops between leguminous hedgerows grown along the contour lines of the hillside. In addition to alleviating soil erosion, these leguminous shrubs also provide "green manure" to maintain soil fertility, livestock fodder, and wood for cooking and construction.

With contour farming, many farmers gradually stopped shifting to new fields every few years and began adding more permanent tree crops such as coffee, cacao, and fruits. By improving agricultural production and farm income, farmers have reduced their dependence on the forest, thereby reducing the threat to their environment. Attitudes are slowly changing, as farmers become aware of the need to protect the environment for their own welfare as well as that of future generations.

The people are also working on ways to continue harvesting the forest for its hidden products such as spices, dyes, hardwood lumber, and oils. Through their own initiative, they are creating "family forests." Year by year, the family forest captures the imagination of upland farmers. The forest is being reconstructed, story by story, year by year, alongside the farmers' fields. One year mahogany is added, in the next cinnamon seedlings are tried; slowly the vines of the famine tubers begin to wrap around the trees of the newly emerging forests ....

How Yayasan Tananua Achieves Results

Start Small, Start Slowly. Farmer activities are based on the needs of the community. The improvement of existing technologies and the process of raising awareness involves the community from the very beginning. During this process, extension workers live with the villagers in order to better understand community needs and potentials and to act as facilitators in the planning and decision-making process. It is important to remember, however, that the extension workers in the program, in almost all cases, are farmers themselves—not outsiders coming in to impose unwelcome ideas.

It is important to start with small-scale experiments with a few farmers, so that results can be monitored and evaluated collectively. The success or failure of the trials will have a significant impact on the attitude and trust of the farmers toward any new information.

Build Upon Model Farms. New ideas are tried on the farms of the extension staff and the farmers themselves, which become places of learning and sharing experiences. The extension staff works more intensively with a few farmers who are pioneers in the process and who have a long-term perspective. This process is significant in planning and developing model farms that can and will be replicated by other farmers. Extension activities will generally succeed if farmers start from their own farm: methods that have already been tried by a few farmers are more easily transfered and accepted by other farmers.

Simple Technology. The selection of a technology is based on its adaptability to local conditions, relative inexpensiveness, ease of use, and appropriateness. Technologies are developed based on farmers' experience and extended only one at a time. This facilitates communication with farmers, who often have limited formal education.

Farmer Leaders. For long-term sustainability of program efforts, Yayasan Tananua is committed to developing leadership skills among farmers who voluntarily take an active role in teaching their neighbors.

Extension Media. The adoption of technologies is facilitated by extension media (simple, heavily illustrated booklets, slides, posters, etc.) that are developed for use at the community level. Extension is then carried out on the farm, in the house, at church, or during village meetings and farmer group discussions. Extension using media is always accompanied by practical demonstrations on the farm. This process continues until the extension staff are convinced that the farmers are able to implement the technologies themselves and spread their knowledge to other farmers.

Quarterly Meetings. Yayasan Tananua's quarterly meetings, which

We should always remember the three basic principles in adopting any soil conservation practices:

1. EROSION CONTROL
2. MAINTENANCE OF SOIL FERTILITY
3. CONTROL AND ABSORPTION OF SURFACE WATER

Using proper soil and water conservation practices will not only save our most precious resource, but will increase our harvests and make our farm more productive and even more beautiful.

A page from the book *Introduction To Soil and Water Conservation Practices.*

are rotated to a different village every three months, are open to all farmers and staff. During these meetings, farmers have an opportunity to share experiences about their activities, discuss their problems, and make plans together for Tananua's future. Farmers also collectively visit and critique each other's farms during participatory evaluations. Farm planning, in which farmers plan the future of their farms and "family forests" by drawing maps and diagrams, is a popular activity of the quarterly meeting. Not surprisingly, these meetings are an important source of farmer solidarity.

The marginal living conditions of farmers force them to consider their immediate needs before they can start planning for the long term.

Cross Visits. Cross visits are organized between farmers from the sàme village, from different villages, or from different islands. These visits are intended to improve the skills and knowledge of the farmers as well as broaden their perspective in further developing their farms. For those farmers living in remote areas, contact with outside villages seems to generate much enthusiasm for farm improvement as well as for experimenting with their own ideas.

Remaining Challenges

Despite having achieved a certain degree of success, Yayasan Tananua is the first to admit that a number of challenges remain. The marginal living conditions of farmers force them to consider their immediate needs before they can start planning for the long term. Hence, motivating farmers to begin with basic soil and water conservation measures as a way of protecting their environment remains an uphill battle with some traditional members of the farming community.

Farmers living on the boundaries of the remaining forests in Sumba are haunted by changing land rights and the thought that any day, they could be resettled far from their traditional lands. A challenge remains for these buffer-zone communities and government officials to work out a plan in which farmers can remain on the land if they use it appropriately, and can be considered custodians rather than enemies of the protected forest.

11

Finding Local Solutions: What it Really Means

Gus Polman

A conventional approach to development is for outsiders to come into communities and initiate projects that require large amounts of money and professional expertise. Even with the best of intentions, many of these projects collapse, often leaving the recipients worse off than when they began. The missing ingredient for the success of these projects is the input of the people who are to be affected—the community members themselves. Fortunately, there has been a fundamental rethinking of the role of nongovernmental organizations (NGOs) in community development. MYRADA is one development agency that has recognized the importance of local knowledge and initiative to the success of its work. This organization, founded in 1968, is involved in rural development in approximately two thousand villages in the southern Indian states of Karnataka, Andhra Pradesh, and Tamil Nadu. The following article looks at how MYRADA uses a technique they call Participatory Rural Appraisal in a grassroots approach to community development. This article was prepared by Gus Polman from HOPE International, a funder and partner in many of MYRADA's projects in India.

Community control is a phrase and philosophy that has guided MYRADA's work in more than forty-two different watershed areas in southern India. MYRADA finds that when nongovernmental organizations work with systems that reflect traditional practices—what people do themselves—they are maintained.

Because of this understanding, MYRADA has developed a teaching and learning technique that relies on the knowledge of people in the community. Formally called Participatory Rural Appraisal (PRA), it is simply participatory planning of natural resource development and management projects—a process in which villagers share from their experience to propose effective courses of action for development of wastelands and watersheds.

PRA is a marked contrast to the way villages are often approached by people lecturing them about what they should do. PRA discovers and employs the abundant and untapped resource of the rural people themselves. It attempts to build on this for sustainable rural development.

Villages selected for the PRA exercise are usually ones where MYRADA or a partner agency has already established a relationship and where development programs are taking place or are being proposed. This ensures that a village will have a concrete avenue of response to needs expressed during the process.

Equalizing the Experts—An Example

Garudu Kempanahally is a small village of fifty-eight families in southern Karnataka. Degradation of forests and soil erosion in the region are linked to poverty, as people are forced to consume the area's resources but get poorer as their resource base dwindles. The village wanted to find a way to manage their resources sustainably. They decided to use the PRA exercise in a three-day community effort in May 1990 to develop an integrated plan for its area watershed. About twenty people from MYRADA's Bangarpet project and some forestry department staff participated with the villagers in exchanging ideas, but the outsiders' involvement was characteristically different from the usual pattern.

The process began with a few "equalizing" exercises, in which all participants did routine village tasks—chopping firewood, digging manure, washing clothes. The outsiders, who were experts in their fields, found that these simple village tasks were not so simple after all.

Then, all took part in a completely visual exercise illustrating the local seasons. They used stones, sticks, and different colored seeds to represent the months, festivals, rainfall patterns, production, work rhythms of men and women, incomes, and expenditures. Participants next diagramed the regional watershed and made a model of it, indicating land and soil types and uses, water management, and treatment plans. The visitors accompanied groups of villagers for observational walks.

The region's upper catchment area is largely deforested, but in a seemingly bare terrain villagers revealed an amazing variety of plants. In a small area about the size of an average room in a Canadian house, a farmer showed twelve different species and the villagers' use for every one of them. There was a kind of flower to stop diarrhea, something to cool the circulatory system; something else to treat minor fractures and sprains in cattle; an aromatic weed used to repel pests from stored grain, and even a berry that indicates the start of the monsoons. For many of the outsiders it was astounding to see how much the people are in harmony with their environment. The villagers shared their expertise in using scarce resources resourcefully.

At a village meeting on the second day, the group decided it was important to reforest the upper catchment in order to minimize the process of soil erosion and silting of the water catchment tank. They considered adopting several approaches for soil conservation: a mechanical approach involving check-dams in waterways and a biological approach with a preferred choice of shrubs and trees. Farmers ranked local trees by diagraming the major species and their uses. This helped to decide the appropriate mix of species to be used in implementing the plan.

The third day focused on village social groups and institutions. One group indicated health needs and problems. Another prepared a time line of local histories by interviewing elder villagers. Some people discussed patterns of change in customs and practices. Another made chapatti-like paper cut-outs of people and institutions. Others drew a pictorial map that diagramed wealth rankings.

Participants wanted to make sure everyone benefited from the social forestry plan. At that time, three of the families were landless. There was also one farmer with 80 acres and another with 67. Villagers were thinking of ways to get one large-scale farmer to sell 5 acres.

Deciding what they needed was just a beginning for the people of Garudu Kempanahally. But for the first time, the people of the village had a chance to shape the future of their habitat according to their own priorities and perceptions.

The Challenges

The technology used for village improvements is quite basic, but the environments and social conditions in which the communities operate don't make for easy application. There have been disappointments with failed wells and inadequate watering and maintenance of planted seedlings and nurseries, due to combinations of unpredictable groundwater resources, lack of rainfall, funding delays, and general lack of experience. There was also serious damage to some nurseries and planting sites from a devastating cyclone in May 1990.

Another threat to this positive approach to local community development is the takeover of land and tree assets by powerful landowners and government, which of course completely undermines the program's achievements. Some measure of protection could be provided by consolidating and legalizing title to the lands and resources of the poor. MYRADA has helped communities and clusters of villages organize to gain control over their resources. And not everyone is antipeople; there are many good people in government and MYRADA's approach is to work with them wherever they are found.

The PRA process has assisted many villages in deciding on a program of healthy, sustainable development in their locales. When communities have a heightened awareness of the incentives to look after their own land and trees, the environment in previously degraded areas promises to improve dramatically.

For more information about MYRADA's Participatory Rural Appraisal process, contact James Mascarenhas of MYRADA, 2 Service Road, Domlur Layout, Bangalore, India 560071, or phone (91) 812 576 166.

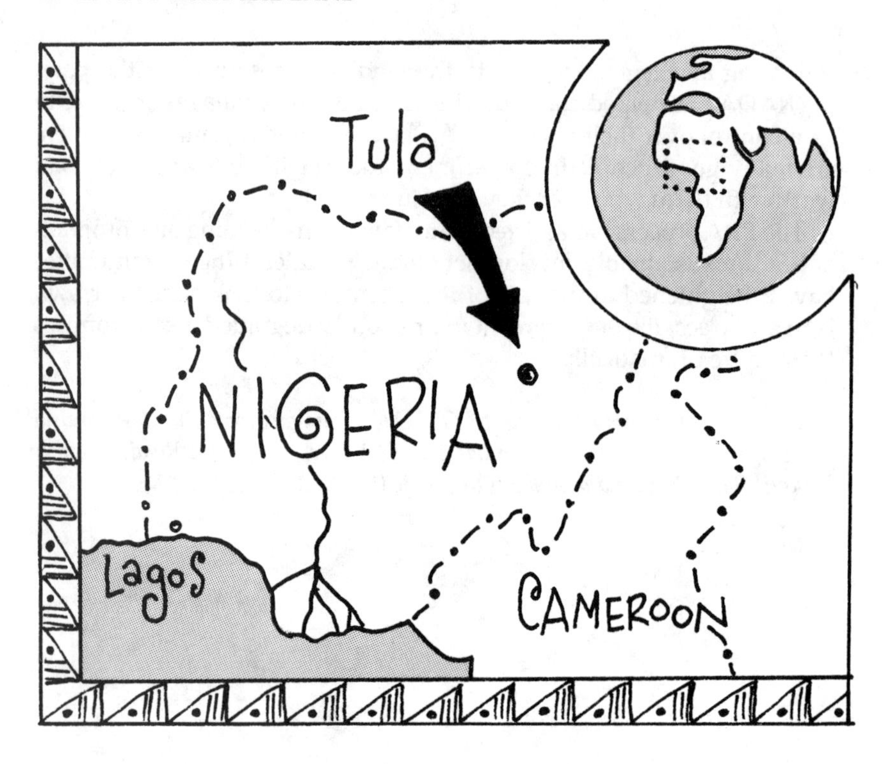

Contact:

Mark Lutes

Faculty of Environmental Studies
York University
4700 Keele Street
North York, Ontario
Canada M3J 1P3

12

Enduring the Wind, Announcing the Rain: Cultural Endurance in Nigeria

Mark Lutes

All too often we flatter ourselves in thinking that the concept of living sustainably is a new idea. Long before the term sustainable development was even coined, communities of people have developed highly advanced ways of working with nature to provide what they need, without destroying their environments. Such is the case in this story, which comes from the Tula region of Bauchi State in Nigeria, where the people have intensively farmed the fragile land for centuries while maintaining and even improving its ability to sustain them. Centuries ago, Tula villagers perfected techniques of stone terracing, organic fertilizing, intercropping, and crop rotation, which have proven their effectiveness in maintaining the agricultural base of the community.

While serving as executive director of the Conservation Council of New Brunswick, Mark Lutes visited the Tula region of Nigeria, where he had a conversation with Timothy Durkun Suyi, the assistant headmaster of the elementary school in Tula Wange. What follows is taken from his notes during that visit. The Conservation Council operates a farm demonstration project named after the Tula region, to reintegrate sustainable agriculture techniques into Canadian farms.

After the fast two-hour drive east from the state capital of Bauchi, the last leg of the trip to Tula was a dramatic change of pace. We turned off the paved highway onto a track that snaked across bare rock and ground for twenty kilometers. Finally we ascended a steep hillside and passed through a natural rock gateway and into the first Tula village. The rainy season, which begins in May, was just ending and the countryside was green and full of life.

I had been told about Tula by a couple in New Brunswick who had traveled through the area in the 1960s while teaching in Nigeria. They concluded that we in Canada could learn a lot about farming sustainably from places like Tula. However, neither their description of the region nor the beauty of the landscape on the drive out had quite prepared me for what I was to discover.

Tula is actually comprised of three villages—Tula Wange, Tula Baule, and Tula Yiri—that are perched on a chain of low mountains surrounded by flat plains and valleys. The villages are about four kilometers apart and have a total population of 33,000. Entering the first village of Tula Wange, the thing that immediately struck me was how lush and green the mountaintop was, even compared to the surrounding countryside. Every square metre that wasn't vertical rock was alive with an amazing variety of plants. Scattered unobtrusively under small groves of trees, along the road and part way up the steep hillsides were collections of round huts. We stopped to look at the first steep hillside we passed, and found that among the millet and guinea corn were horizontal rows of stones. These stones were a foot or so high and five to ten feet apart, and were placed all the way up the hillside.

Arriving at the center of Tula Wange, my guide and interpreter, Ahmed, asked where we could find someone to tell us about the place. We were directed to a nearby primary school. There we introduced ourselves to Timothy Durkun Suyi, the assistant headmaster of the school, who had a thorough knowledge of the agricultural techniques employed in Tula. He invited us into his office, where we were joined by some of the school's other teachers, mostly women. They proceeded to give a detailed description of the farming practices, occasionally discussing the finer points among themselves in the Hausa language.

The Sustaining of Generations

Tula has existed in the same location for many centuries and the distinctive agricultural practices were perfected early in the communities' history. The villagers stayed at the site partly because the high altitude ensured fairly cool temperatures all year round and also because the steep cliffs and limited access to the top made the site very defensible in times of unrest.

Most of the farming on the hilltops close to the villages is carried out by the women, and this is where the most distinctive and innovative farming techniques are found. Some time before the rainy season, the women rebuild the stone terraces across the slopes, repairing the damage

Holding Back The Winds: Millet and cocoa yams planted on large stone terraces on a steep hillside. The leaves have been buried around the yams to help them decompose, and provide nutrients.

Photo: Mark Lutes.

done by grazing livestock. The terracing helps to retain the precious soil on the steep slopes, which otherwise would be washed away by the heavy rains and carried off by the winds. This area is quite well known for the devastating action of the winds during thunderstorms.

The ideal time to sow most crops is just before the rains come. At the appropriate time certain elders in the village make predictions, based on their experience and signs of weather patterns, for when the first rain will fall. They tell the women when they should sow the crops. Then begins the complex system of intercropping, crop rotation, and organic fertilizing.

Cocoa yams are one of the primary staple crops grown on the slopes, along with maize, millet, and guinea corn. Unlike most crops, cocoa yams are always planted after the first rain. They demand a lot of nutrients, and extra care is taken when they are planted. Before the rains, animal manure is brought to the land and piled in heaps, along with dried grasses that will decay over time. After the rain the women till the ground, and the heaps of manure are spread and mixed in. After tilling and mixing, the cocoa yam is planted and then the dried grass is spread out as mulch to cover the soil.

At the edge of the terraces, maize and guinea corn are planted. After germination the women thin the corn and maize, leaving two maize stems and two to three guinea corn per bunch. New grass comes up and must be weeded. This is done by hand and sometimes the land is tilled with a hoe under the dried grass. When the cocoa yam is almost ripe and about two feet high, the women bury the dried grass in the ground under the cocoa yam, without disturbing the roots. This helps it to finish growing and keeps the land fertile, so that next year they can plant guinea corn. Guinea corn can then be planted for two years and the next year millet is planted. In some areas the cocoa yam is planted two years in a row. These areas are called *mar* in the Tula language, meaning they grow the best cocoa yams. Groundnuts are also planted periodically to put nitrogen back into the soil.

Any break in this cycle of intensive agriculture would mean the permanent loss of the soil that the women before them had so patiently tended over the centuries.

Without the continual use of these methods for maintaining the soil and its fertility, agriculture around the village would rapidly become impossible. The heavy rains and strong winds would soon carry all the soil off the steep hillsides and leave little but bare rock. Also, given the speed with which organic materials decompose in tropical soils, a constant infusion of mulch and organic fertilizers is needed to maintain soil fertility. Only the steady and meticulous attention over the centuries could have maintained the soil in its present condition. Any break in this cycle of intensive agriculture would mean the permanent loss of the soil that the women before them had so patiently tended over the centuries.

The men farm the flat lands two to ten kilometers from the village. These areas are known as the far lands. There they use traditional techniques, including oxen and large-scale farming. Cotton, groundnuts, bambara nuts, guinea corn, maize, millet and beans are raised there as well as rice, which is grown near the river banks. Some men tried artificial fertilizers, but rejected them because the land became weaker as a result. They didn't hold the soil together and keep it moist like organic fertilizer and manure. Rain leaches the artificial fertilizer down into the ground where only the deep-rooted crops can reach it. Artificial fertilizer was also found to contaminate the drinking-water supplies. Manure is not used much on the far lands because the roads are too poor to transport it, unless there are villages near. On the far lands, when the soil gets poor, it is allowed to be fallow for two to three years to let it rest.

What We Can Learn

On the drive back to Bauchi I thought about how much Tula had that was missing from the world to which I would soon return. Industrialization, consumer culture, and global markets are reducing communities to little more than aggregates of people buying and selling commodities. In the process, nature is reduced to having value only if it can provide an immediate economic payoff. The Tula community seemed to have a sense of self- reliance, a rich store of wisdom and traditions, a respect for the needs of future generations, and an integration with the environment—qualities that are quickly disappearing in much of the world. My experience at Tula provided a brief but exciting glimpse of what a community can be.

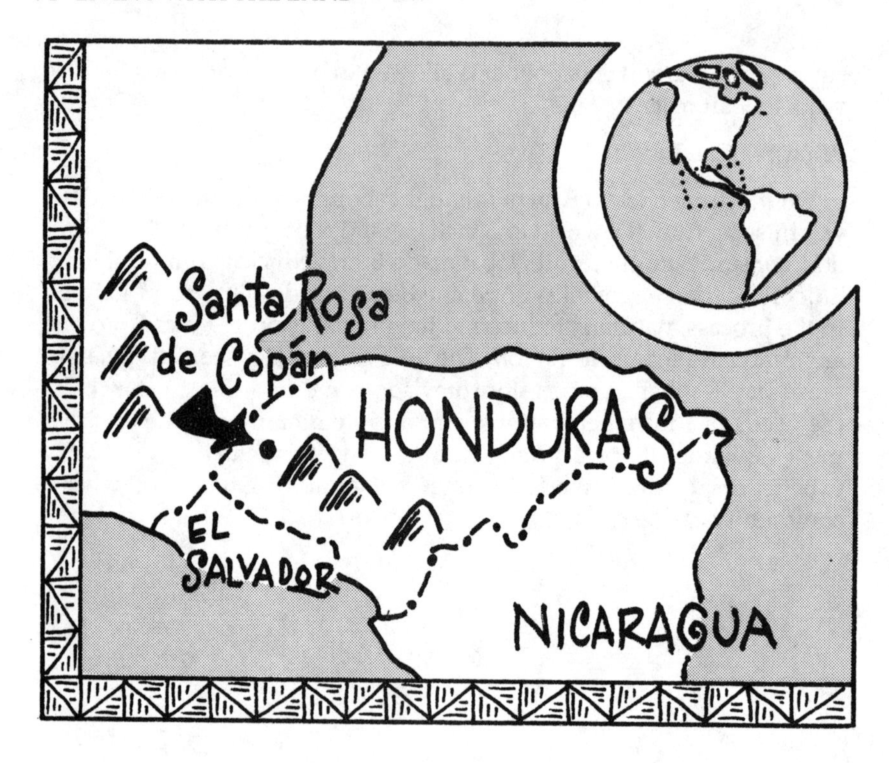

Contact:

INEHSCO
3 Calle S.E., Barrio Santa Teresa
Santa Rosa de Copan
Honduras

tel: (504) 62-0826 or (504) 62-0755

13

Healing by Tradition: Herbs and Liberation Theology in Honduras

Philip Tamminga

From the philosophy that people must be the creators of their own development has grown a small grassroots organization dedicated to empowering the rural population in Honduras. The Honduran Ecumenical Community Services Institute (INEHSCO) focuses on health as a way to involve people actively in their own development. The group discovered that people know a great deal about improving their own health without the use of expensive drugs from outside development agencies. INEHSCO began opening health centers where people could find traditional local remedies, as well as producing educational radio programs about how to grow organic food, medicinal plants, and other topics. INEHSCO is putting the theology of liberation into action—liberating people to help themselves.

Canadian Philip Tamminga and his Honduran wife, Marlen, are both involved with INEHSCO. They share a strong interest and background in organic agriculture as well as grassroots political activism. Tamminga is working with INEHSCO to produce educational radio programs and to establish a radio training and production center.

During a recent visit to Honduras, I had the opportunity to speak with Father Fausto Milla, a Catholic priest in the town of Santa Rosa de Copan. Milla is the general coordinator of INEHSCO, the Honduran Ecumenical Community Services Institute, an organization committed to improving the lives of Honduras' rural campesino population through grassroots projects and popular education, mainly in the field of health and nutrition.

Our first conversation took place at the Tienda Naturalista #1, one of two natural food stores run by INEHSCO in Santa Rosa. As I listened to Fausto talk about INEHSCO's activities, I began to understand why members of the Honduran military continue to keep a discreet watch on the store. In Honduras, the promotion of natural medicines and traditional healing is not just a new-age health fad; rather it is work that directly confronts the relationship between health conditions in the countryside and the distribution of wealth and power in Honduras. The institute's work is seen by some to be a subversive alternative to the status quo.

In Honduras, the promotion of natural medicines and traditional healing is not just a new-age health fad.

INEHSCO was formed in late 1980 by a group of Hondurans concerned about the continued and systematic neglect of Honduras' rural population by the national government and international aid agencies. The group brings together a wealth of experience from the campesino (peasant) movement, popular organizations and church-based assistance programs and was inspired by the teachings of liberation theology. Liberation theology is a term coined during the 1960s by Catholics concerned about the church's seeming indifference toward the poverty and social injustices committed against Latin America's poor. Liberation theology calls for a more democratic church as well as social and economic justice through local community activism. This philosophy guides INEHSCO's activities in communities. Building on the existing infrastructure of the local parish, the institute tries to reach as many members of the community as possible with its educational activities.

Since its inception, INEHSCO has been committed to the empowerment of people for control of their own development. The organization is critical of development and assistance programs that in fact only help to perpetuate the system of economic dependency and, worst of all, the deep-rooted sense of inferiority that goes along with it. On the surface, large-scale assistance projects such as rural electrification may seem to provide an improvement in living standards, but often there is not adequate consideration of the cultural and environmental destruction caused by these projects.

"I think many of these projects rob campesinos of their own initiative and integrate them more fully into the economic system that is respon-

Pueblo Sano Y Nutrido (A Healthy and Nourished People): Father Fausto Milla sees this as the key to development. In the Tienda Naturalista No. 2, they sell traditional herbal medicines as well as locally-grown organic fruit and vegetables.

Photo: Phil Tamminga.

sible for their poverty in the first place," says Fausto Milla. "And because they are designed and implemented with little or no participation from the local population, the real needs of the community are rarely addressed."

By working directly *with* campesinos, the institute soon recognized that poor health conditions are the biggest concerns of the rural population. Medical infrastructures are virtually nonexistent, and medicines imported from the large transnational drug companies are in short supply and priced beyond the means of most campesinos. The problem is compounded by a poor diet and unsanitary living conditions, as campesinos are forced off the best agricultural land by powerful cattle ranchers and growers of cash crops like melons for the export market. INEHSCO is committed to improving the standard of health in the region through grassroots projects. This approach has gained the institute widespread respect and participation from the local population. As Fausto Milla explains, teaching illiterate campesinos to read is of little benefit to them when they are hungry and malnourished everyday.

Because the oppressive economic system denies the rural population access to medical attention, INEHSCO's solution has been to revive popular natural remedies that are readily available at little or no cost to all campesinos. Plants like malanga, passion flower, and wormseed, which were once considered weeds, are now being cultivated in the home for their medicinal and nutritional value.

In 1985, the institute opened a small natural food store in Santa Rosa de Copan and began giving classes in traditional herbal remedies. The response was overwhelming, and a second store soon followed. More important, though, was INEHSCO's community outreach program. Campesinos were trained in all aspects of natural health care, after which they returned to their communities to share the information with others. In one community, a government-run health clinic reported a 70 percent

A Song of Health

In November 1990, the Honduran Ecumenical Community Services Institute (INEHSCO) hosted the Fourth International Congress on Traditional Medicines in Santa Rosa de Copan. The congress brought together international experts on natural medicines as well as rural farmers (campesinos) whose knowledge of traditional healing was based on years of practical experience. In many cases, the campesinos were the ones teaching the experts about the properties of plants being studied in laboratories. In conjunction with the congress, the institute sponsored a song festival called For Health and Life. Campesino groups were asked to compose songs that could be used to spread the message of natural healing to other campesinos. Below is a translation of one of the songs performed at the festival.

Plant Extracts

(from the community of Siete Cuchillas)

Friends, I'm going to tell you about
something that has grown out of this region.
It's a health program to help everyone.

There's no more long line-ups
in the health centers or hospitals
because we are curing ourselves now
using natural plants.

decrease in admissions after INEHSCO began activities in the area.

To reach a wider audience, INEHSCO began producing educational radio programs that provide information on how to grow medicinal plants, sanitation, diet, organic gardening, and a host of other topics. Today radio stations around the country broadcast INEHSCO's programs to an estimated 10,000 to 15,000 listeners, some as far away as El Salvador and Guatemala.

But INEHSCO's impact can be best seen close to home, in Santa Rosa. While Fausto and I were speaking, a young boy knocked repeatedly at the door of the store, which was closed at the time. Instead of spending his money on a cola ("the black waters of imperialism," as Fausto Milla

I don't get worried now when I get sick
with natural plant extracts
I can cure whatever's wrong.

We thank heaven, and the Institute too
for bringing us this knowledge
as a demonstration of love.

And if there's someone who opposes
this humble mission
It's because they're far removed
from our sad situation.

Today, things are so expensive
and the poor can't make ends meet
but hand in hand, united we'll stand
and face the struggle.

calls it), he had come to the store to buy a natural fruit-juice drink, because it "helped him to stay healthy and strong."

The most exciting thing about INEHSCO's programs is that all of these traditional cures and medicines come from the campesinos themselves, revived from the memories of the elderly and shared once again with the young. One little girl summed it up when she said, "My grandmother says that anyone who has a lemon tree has a pharmacy in their house!"

All of these traditional cures and medicines come from the campesinos themselves, revived from the memories of the elderly and shared once again with the young.

For the campesinos, the knowledge that they are capable of transforming their own lives has had a profound effect, and has led many to question the economic, political, and social systems that have kept them in poverty. For this reason, the Honduran military and elites are openly hostile to INEHSCO's efforts to promote social change. Fausto Milla was expelled from the country for four years, and INEHSCO workers are often targeted with intimidation and threats.

But Fausto Milla remains confident in a better future for Honduras. "For five hundred years since the Spanish conquest, we have been told that we are ignorant Indians and that Europe or North America had the solutions to all our problems," says Fausto Milla. "Now we are rediscovering our own knowledge and traditions, the same ones that helped to build the great Mayan civilization, and this will help us to transform this society and create a new one based on justice, peace, and love."

Part Four

FOLKS IN CITIES AND TOWNS

Most of us could name communities that once had a special quality about them, or the potential to achieve it, but which have since become part of a larger, sprawling metropolis. All too many of our most favored communities and landscapes have or are about to become paved over for strips of highway, franchise architecture, and the like, trading in a sense of place and authenticity for a feeling of sameness and homogenization .... The enormous pressures for growth and development and the resulting effects on the landscape have seemed specially overwhelming in the last decade or so .... We urgently need to find more effective strategies that will enable our communities to grow in ways that enhance, not degrade, the qualities that lend them distinction and character.

—Michael Mantell, *Creating Successful Communities: A Guidebook to Growth Management Strategies*

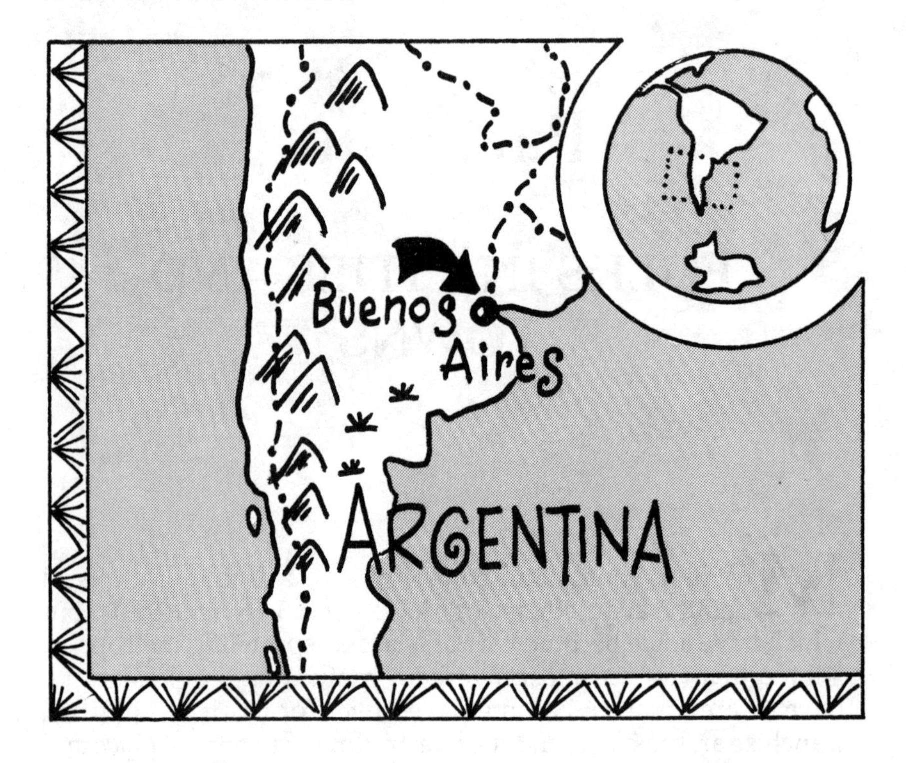

Contact:

Ana Hardoy

International Institute for Environment and Development America Latina (IIED-AL)
Corrientes 2835 - 6 piso B - Cuerpo A
(1193) Buenos Aires, Argentina

tel: (541) 961 3050; fax: (541) 961 1854

14

Overcoming Barriers in the Barrio: Organizing in Barrio San Jorge

Lorenza, Quela, Estela, Susana, Brenda, Margarita, and Ignacio, all members of the Administrative Council for Barrio San Jorge; also Ana Hardoy and Jorge E. Hardoy, with Ricardo Schusterman, IIED-AL

Living on the margins of the city, on land and in structures they do not own, squatter communities are often the last in line for even basic necessities. Not only is land tenure insecure, but low-income communities face immense uncertainties about their future income, nourishment, and survival.

Individuals and organizations need to learn new ways of working with these communities to improve their quality of life. Since 1987 the International Institute for Environment and Development America Latina (IIED-AL) has been involved in the settlement known as Barrio San Jorge, on the outskirts of Buenos Aires, Argentina. They are confirming that the best way for an assistance group to work in a community is by listening and learning from the people who live there.

This is the story of Barrio San Jorge, as told by the people who live there as well as by members of IIED-AL. The staff describes their goals when working with squatter settlements and shares the lessons they have learned from San Jorge. Meanwhile, the neighbors share their perceptions of their own growth. Despite being a group of people forced together by political and economic circumstance, they are learning to get along—creating community out of the ruins of the streets.

The Residents: "A Difficult Beginning"

Our neighborhood or barrio is located thirty-five kilometers northwest of the city of Buenos Aires in Argentina. On one side is the Reconquista River and on the other is Highway Route 202. The population here is currently 630 families, some 2950 inhabitants. In the beginning, the land we occupied we did not own.

Making improvements in our barrio has taken a long time. Different people have been coming in to help with things like getting electricity and water. At the beginning a large tank was constructed, which the fire department periodically filled. When there was no water, we'd carry canteens to fill with water from our workplaces. With help from the church and some private groups, we made several attempts to improve the neighborhood, including finding supplies of water and electricity, fixing the streets, organizing garbage collection, and digging ditches to drain the water. But many of these initiatives did not last long.

The old part of the barrio was created in 1962. In 1980, the municipal government of San Fernando developed a new sector in the neighborhood, with people from different areas in the region. There were many problems between the new and old sectors—fights, gangs, and confrontations of all kinds. The barrio was a dangerous place. You could not walk around at night because you'd be knocked down or hurt. Such problems used to occur on a daily basis. But there were many working families in the barrio who wanted, and still want, things to improve.

Since the recent arrival of a team made up of municipal workers, technicians, and others from private, nonprofit institutions, there has been more commitment to improving the conditions of the barrio. Neighbors have even started to help one another.

Coming Together

We see the building of the Mother and Child Center as a significant step in bringing us closer together. In the beginning, many single women had no childcare support during the day. We changed this situation with the help of CARITAS, a group of private institutions, and some of the women and men from the barrio. It is very important for women to be able to work and support themselves and to know that their children are being cared for. The children at the center are from two months to four years in age and are cared for from 8 a.m. to 4 p.m. daily. At the center they are given breakfast, lunch, and snacks, as well as health and hygiene care.

A group of mothers also wanted to have better work opportunities.

They got together at the beginning of 1989 and invited other mothers to participate in the organization of a sewing workshop. The mothers, with the help of professionals who were already working in the Mother and Child Center, applied for support from the German embassy. With this money they bought three sewing machines. This is how the training program for the mothers of San Jorge began.

Today, in spite of our previous differences, things are already improving and there are no more confrontations between the new and old sectors of the neighborhood.

This same group of women who called themselves Mujeres Unidas (United Women) organized other tasks as well. They functioned then as our Neighborhood House does now. A variety of activities takes place at the Neighborhood House today, including a workshop for expressing oneself, a sewing workshop, and a football team. These events are organized for young people between ages 9 and 14. At the house we have psychologists, social workers, and some mothers and fathers. In the same house we have a telephone that serves the 630 families of the barrio.

In the future the house will be used for many other activities. It is the largest space in the barrio and is good for parties. It is also the place where the neighborhood association meets once a week.

Today, in spite of our previous differences, things are already improving and there are no more confrontations between the new and old sectors of the neighborhood. There is more tolerance, more friendship, more cordiality.

In the Wake of Paternalism: The IIED-AL

When the head of our assistance group (IIED-AL) began construction of the Mother and Child centre in September 1987, only sixteen people in San Jorge (mostly women) were interested in community activities. The great majority of the population looked upon the construction of the center with skepticism. Calls for help were ignored and frequent jokes or negative comments were made. Such an attitude was understandable, since the community had seen the comings and goings of several different outside groups that worked without involving the residents.

As the construction of the Mother and Child Center developed, so too did interest from small groups of residents who supported the initiative—again, mostly women. Inevitably, some neighbors raised the question, "What are we going to do next?" But the vast majority of inhabitants adopted a negative attitude. They said nothing could be done because others did not want to cooperate. It was assumed that everyone who launched an initiative to raise funds would keep the money for themselves.

This attitude contrasted with the generosity most inhabitants showed toward each other. Despite their poverty, they willingly purchased tickets in a neighborhood lottery organized by friends, or helped a neighbor who needed financial support to bury a relative, celebrate a birthday, or purchase a long-distance bus or train ticket. However, these attitudes changed if the initiative came from an organized group. Mistrust of such initiatives still prevails.

How can these attitudes be explained? Since the 1940s many of the urban poor and unemployed in Argentina have expected the state to answer their need for food, clothes, blankets, school books, and the like. Argentina has never had social insurance for the unemployed. Good connections with political bosses and committees has long been the way to solve problems. This attitude was encouraged and even promoted by the state as well as the church and some private charities. This "assistance" orientation, and the paternalistic attitude from those with political power and access to funds, have limited the support available to residents of squatter settlements.

Each low-income barrio—whether a squatter settlement, an illegal subdivision, or indeed a legal development—has its own history. This proves to be an often-powerful force in shaping the attitudes of the inhabitants. For Barrio San Jorge, the history since the late 1960s is closely linked to the activities of a priest from the church of Antioquia, whose attitude toward the inhabitants, sometimes positive but frequently negative, gradually undermined any attempt by the community to organize. This priest was a strong and despotic personality who always knew what was best for the barrio. He never accepted the opinion of the inhabitants. He even had the barrio named after him, for he was called Father Jorge.

Father Jorge's activities in the barrio extended through the 1970s. During seven of those ten years, the government of Argentina was controlled by non-elected military officials and all political and technical positions at the national, provincial, and local levels were appointed by the armed forces. The priest had close links with the military and the government. Conscious of his political support, Father Jorge ran the barrio like a military camp and even imposed a nighttime curfew.

In those years the barrio was smaller, consisting of what is now locally called the *barrio viego* (old barrio). The new part of San Jorge was formed in 1978, as new families arrived from among the hundreds who were forcibly evicted by the military government, who wanted the land where the squatters had previously lived.

Perhaps this history is one way of understanding how power has been used in Argentina, and one way of explaining the chronic mistrust that the residents of San Jorge have for outsiders, for their proposals and interventions.

A Better Approach

When we first meet with residents to discuss a new initiative, they all seem to agree with the proposals. If we ask residents what they think, there are no replies. Time is needed to allow new ideas to mature, and it is only in a second or third round of discussions that individuals in the community make comments or suggestions. Some good initiatives have been aborted because they were presented by someone from an outside agency. Failures are almost inevitable when outsiders try to impose their points of view.

The inhabitants of both the old and new parts of the barrio have been used and manipulated by political and religious leaders .... They never express their political views openly.

Most external agencies also fail to understand that each community is different. For each one, these differences are shaped by their particular history and the specifics of their social and natural environments. In the case of San Jorge, the inhabitants of both the old and new parts of the barrio have been used and manipulated by political and religious leaders. They have been threatened, expelled from another site, and frequently subjected to harassment. They never express their political views openly. Nor do they take positions about their lives under democratic or dictatorial regimes.

The team from IIED-AL has never asked the inhabitants for anything: not votes, not attendance at political rallies, not attendance at church services. This has been something unusual for the inhabitants. The IIED-AL team has also had to develop a capacity to listen. Work is continuous but at the service of the community.

The Barrio San Jorge program has taught us several lessons. First, there must be a change in attitude away from the paternalistic approach of government agencies, political parties, and some charities, which reduces the population of a squatter settlement to simple recipients of public charity. This means overcoming the distrust that many government agencies and politicians have of community organizations. Instead of trying to capture the leadership of incipient community organizations, political parties can better serve the popular cause by letting them freely elect their own representatives and develop their own activities.

The most important lesson we learned was the need for permanence. The continuous presence of the team in the barrio has been fundamental. And, as each community is different, the only way to help their members is to listen to the voices of the barrio, both the individual and collective voices.

The Residents: Creating Community

The municipal technicians proposed that we hold elections in order to get ownership of the land we live on. They helped us form a commission of delegates representing the old and new sectors of the barrio, and we started to work together more effectively.

On July 28, 1990, the Government of the Province of Buenos Aires, the municipality, and IIED-AL signed an agreement that promised ownership of the land in the San Jorge neighborhood to us—the people who live here. In order to be represented in the agreement, we organized ourselves into a cooperative neighborhood association, which was the best way for us to work together. Today this association is called the Administrative Council, and it deals with the legal requirements necessary to be part of this agreement.

The first job we took on was organizing a census of the population, in order to prevent the addition of more families to the project. The existing plots of land are few and many families already live on them. The council also follows the process of transfering ownership of the land, to make sure it is given to neighborhood residents. Today we have a certain amount of control over the movements of residents within the neighborhood and have a registry that keeps track of the changing ownership of neighborhood houses.

Through the council and the participation of the municipality, we've had proper garbage containers constructed and have organized garbage collection within the neighborhood. Through the participation of residents from each block within the neighborhood, we have improved the streets with a variety of materials. We have also cleared drainage ditches, leveled streets to ensure safe access for garbage trucks, and installed street lights throughout the neighborhood.

Our Administrative Council has met with other such councils, representing neighborhoods in the region that have similar problems to ours. Together we have formed an Inter-Neighborhood Council for the Municipality of San Fernando, which works with a team of technicians

Paving The Way: "Other achievements include clearing the drainage ditches and improving the streets."

Photo: Isabel Hardoy.

from the municipality. We meet every Thursday to discuss the different problems in each neighborhood, and try to find the most direct ways of solving them.

Looking back, we realize we are much better off now, and we think an even better future lies ahead if we continue with the dedication and enthusiasm we have shown so far.

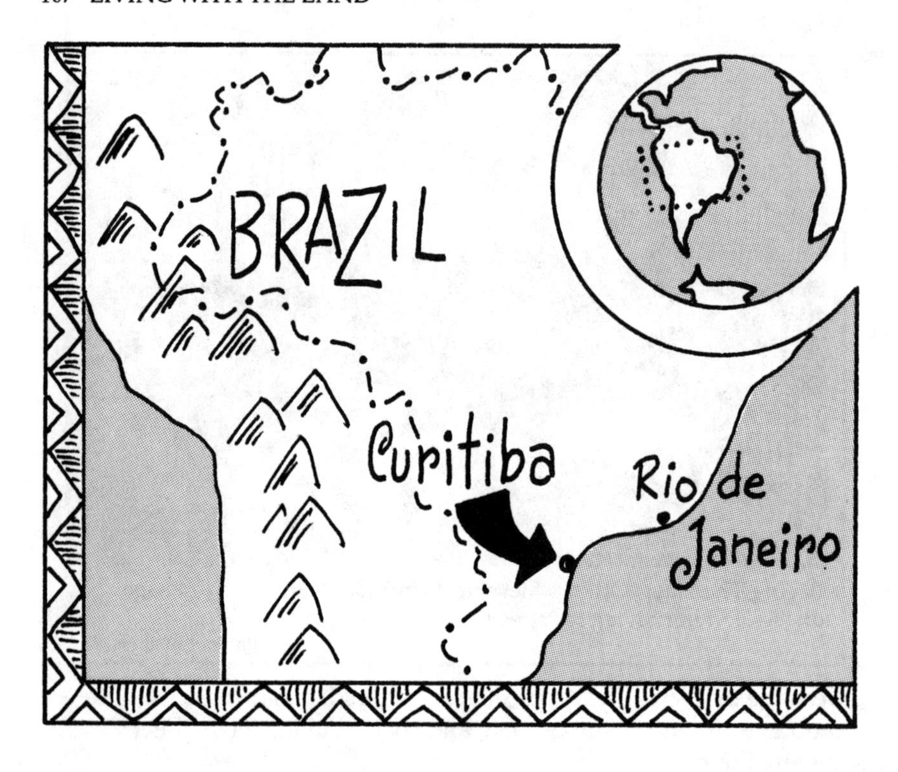

Contact:

Jonas Rabinovitch

Instituto de Pesquisa e Planejamento Urbano de Curitiba (IPPUC)
Rua Bom Jesus 669
Curitiba, Brazil

tel: 55 (041) 352 1414; fax: 55 (041) 252 6679

15

Restoring the City: Curitiba, the Ecological Capital of Brazil

Jonas Rabinovitch

Greening the cities involves more than planting trees and creating new parks. City planners in Curitiba, Brazil are discovering that whole communities need to be involved in developing sustainability in urban areas.

The approximately 1.6 million people of Curitiba want their city recognized as the "Ecological Capital" of Brazil. The process began almost twenty-five years ago, when city planners decided to reconsider their city's growth pattern. They came up with a plan that associated public transit with land use and the road network. What was initially a plan for increased mass transportation has grown into a wholistic management system, with initiatives like a campaign in which recyclable material is exchanged for transit vouchers, a complete bicycle path system, city-wide environmental education, and other programs.

In the following chapter, the city is described by urban planner Jonas Rabinovitch. He suggests the case of Curitiba represents an interesting transition from top-down municipal planning to a more inclusive model in which the whole population participates in environmental initiatives.

The initiative really started when Jaime Lerner was elected mayor of Curitiba in 1971. He was only 34 years old and now confesses that at the time he did not know anything about certain administrative issues. He formed a team of very young people and they were looking forward to doing something for the environment. They began implementing an urban plan that was commissioned back in 1965, a plan that suggested the city concentrate on public transport. So in a time when most Brazilian cities were building viaducts and motorways, Curitiba was giving total priority to decreasing the amount of vehicles. Many roads were designated for the sole use of collective vehicles.

For the first time in Brazil, the central area of a city was returned to pedestrians. As well, the city began increasing the ratio of green areas to inhabitants. In 1970 there were 0.5 square meters per person, and now it's 50 square meters per person. Many places became parks and special "leisure ways" were built. One and a half million trees were planted over the last twenty years.

Creating sustainability in Curitiba has been a process with many phases. The first was mainly on a physical level; afterwards, with time, we came to the conclusion that the process shouldn't be only a top-down one. We are now at the stage of trying to involve the population as much as possible in every initiative.

Garbage That Isn't Garbage

We have a program called Garbage That Isn't Garbage. People learn that there is garbage in the home that is garbage and there is garbage in the home that is not garbage. Communities are now separating all their garbage and there is a special collection van for the recyclable material. According to a recent survey we've done in Curitiba, 78 to 91 percent of the population is separating garbage in the home. The program has been operating for two years and we have already recovered out almost 7000 tons of recyclable materials (paper, plastic, metal, glass) that are reused in some way or another. We've noted that the paper recycling alone spares some 1200 trees daily. For the city, the small financial profit made is reinvested in social programs. But the most important result of this program is that the landfill is being spared. We would have had a problem with our landfill if we had not started this program.

So in a time when most Brazilian cities were building viaducts and motorways, Curitiba was giving total priority to decreasing the amount of vehicles.

The Garbage That Isn't Garbage collection vans take the recyclable material to the Rural Foundation for Education and Social Integration, a municipal institution located in a farm-like environment. There some two hundred people live and work at a simple plant that separates and assembles the materials so they can be sold to private companies that do the recycling. The revenue is then reinvested in the foundation, which is totally self-financed. Any surplus is invested in day-care centers and other social programs.

There is another program called The Garbage Purchase. In certain lower-income areas where the population has settled in river- bottom valleys, there is no access for the garbage-collecting vans. So we made some calculations and came to the conclusion that the same price that we pay to a garbage collecting company could be paid instead to the people who live in the area. We don't pay them with money, but with transport vouchers. This type of exchange is organized in each community. Now more than 20,000 families are involved in over forty-eight communities. Every two or three weeks another community becomes involved in the process, and so far 600,000 vouchers have been exchanged.

When we pay for the garbage with transport vouchers, we allocate 10 percent of what we're giving that community to the local neighborhood association, because they manage the process. We don't believe in paternalism; we don't think we should just *give* money to people. We see it as paying them something like an administrative fee for the work they are doing for their own community. The whole process is managed by the neighborhood. They're the ones who organize the queues and show people where to go, they're the ones who collect the recyclables and put them in proper containers and do the exchange.

We know we cannot tell people what is best for their own street. They know it better than we do. So it's a question actually of giving responsibility to people, so they feel that they are owners of the process as well.

We also just started exchanging garbage for agricultural produce. There has been a surplus of produce in Parana state. So instead of throwing the surplus production in the river or burning it, the producers have agreed to sell it at a reduced price to certain supermarkets. We have agreements with those supermarkets to take this produce to communities and we exchange carrots, cabbages, and other vegetables for garbage. Each bag of garbage is worth a 4-kilogram bag of vegetables. So far about six hundred tons of vegetables have been exchanged.

The Weight of One Person Saves One Tree

There are other ways to participate. For example, we have the Polish Woods, which were actually opened by the Pope when he went to

Curitiba; they are preserved by the local Polish community. We have an Italian district, which is preserved by the Italian community. In a way these areas belong more to the communities than to the municipality.

We know we cannot tell people what is best for their own street. They know it better than we do. So it's a question actually of giving responsibility to people, so they feel that they are owners of the process as well. Our trend presently is very much one of participation, mainly through environmental education. We are not happy just to recycle garbage, because we need to recycle mentalities as well. Changing attitudes is the only way we can have a better future. We find that children are very strong allies in this process. In Curitiba environmental education is not an isolated discipline, but is inserted into the curriculum. In the schools they learn that 2 + 2 = 4, but they also learn that 50 kilograms of recycled paper saves one tree. It's a very simple notion—the weight of a person in paper saves one tree.

There is also the Free University for the Environment. It is not a formal university, but a set of short courses designed for homemakers, taxi drivers, shopkeepers, journalists, teachers, opinion makers in general. People go there and learn very quickly in a practical way how to make a house energy efficient, for instance, or how to separate garbage and the importance of doing that. We have the Green Guard as well, which are the police who work in the parks and are trained to teach people about the environment.

The question of scale is a very important one. We have to balance city-wide initiatives with the goal of involving people at the local community level.

Making Curitiba ecological is not yet a perfect experience. There's still a lot to be done. But the key to our success so far is the fact that we always search for the simplest possible solution and we always try to work with local people—people who know the city, people who like the city. The municipal team is normally made up of very young people, and we try not to complicate the solutions too much. It's also important to stress that it should be done on the broadest scale possible. Mayor Lerner often says that if we do it with only one neighborhood as a test, people won't take it seriously. So we do everything with the city as a whole. But the question of scale is a very important one. We have to balance city-wide initiatives with the goal of involving people at the local community level. I think local residents should take initiatives and should act and feel responsible, while the city administration oversees everything.

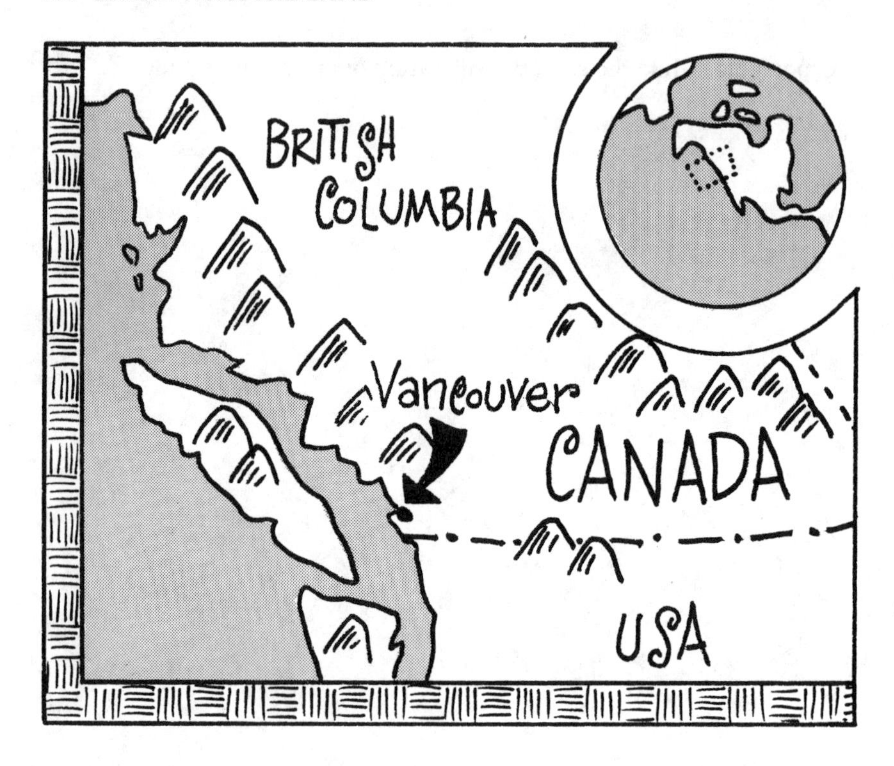

Contacts:

Joanne Hochu
Muggs Sigurgeirson

Strathcona Community Centre
601 Keefer Street
Vancouver, British Columbia
Canada V6A 3V8

tel: (604) 253-3384 or 253-4718

16

Gardening for a Change: Strathcona CommunityGarden

Strathcona Community Gardeners

The inner city—a city within a city—is a term often used to describe places where population density, crime rates, and poverty are high and the standard of housing and incomes are low. In Vancouver, British Columbia the inner city is the downtown east side, where the population consists of single people living in low-rent hotel rooms or on the streets, immigrant families crowded into homes, and others who may not have a choice of where to live due to economics.

The statistics on living conditions in inner cities may paint a bleak picture of gray streets and buildings piled one on top of the other. What the statistics don't show is how the people in these areas manage to create a sense of community. In the Strathcona area (population 5000), located in Vancouver's inner city, members of the community are working together, turning a gray scene to green.

Since 1985, people from the Strathcona neighborhood have been planting seeds in connecting beds on a three-acre piece of land—an oasis among warehouses and busy streets. Growing among the vegetables and sunflowers is a complex, active, and political community, empowering itself and others through the initial act of planting seeds.

When you enter the Strathcona Community Garden from a busy main road, it's like stepping into the countryside. You are met with the sound of birds and an unimpeded view of wild growth stretching across several city blocks. But as soon as you begin speaking with the gardeners and wandering the paths that connect the checkerboard of individual plots, you learn there is a pattern to this garden that goes beyond initial impressions.

There is an orchard and herb garden tucked away in one section, four hundred individual and collective plots for gardeners and local community centers, a compost, a children's play area, beehives, and an untouched wild area and marshland. All of these sections are managed by volunteer committees with interests in particular aspects of the garden. These interests vary from preserving varieties of apples from the seventeenth century, to protecting the marshland for nesting birds, to maintaining the natural growth in the area. People do not come to the garden just to grow their own food, but to become involved in the wider social and environmental concerns of the community.

Social Growing

If you visit the garden, you may meet Juan, a long-time resident of the area who hails originally from the Basque region of Spain. Often Juan can be found turning the soil in a plot close to his own, carefully tended area. This particular plot belongs to the Carnegie Community Center located in the downtown east side of the city. The center uses the vegetables to make reasonably priced, healthy meals for the hundreds of street people who come there every day. Juan says he likes to work in the Carnegie plot now and then to help them out. In fact he seems to enjoy many of the social aspects of the garden, like talking with visitors as he carries on with his work, chopping up yellowing leaves and stalks for the compost.

People do not come to the garden just to grow their own food, but to become involved in the wider social and environmental concerns of the community.

Composting is actually the way the community garden began. For sixty years this area had been used as an industrial dump site. Finally the city banned dumping, but did nothing to clean up the mess. Residents of the area had been eyeing the land as the perfect space for a garden, and in 1985 they approached the parks board with their idea. The board agreed to a one-year, no-cost lease, later extended to five years. Once the lease was secured, the community began the process of developing a better soil base for growing food. They had the original soil tested for metals and, finding nothing harmful, began to reclaim the land, garden bed by garden bed, enriching the soil with decaying leaves and food.

Composting continues to connect the garden with the broader com-

Before: The dumping grounds that became the Strathcona Community Gardens.

Photo: Ellie Epp.

munity. Members have created a space for local organic restaurants and centers like Carnegie to return their kitchen waste to the soil. The garden has been able to accept about one ton of compost each week. Muggs Sigurgeirson says the goal is now to pioneer this kind of community composting on a larger scale, to make it available to those who do not have a backyard. This is the premise of the garden itself, to provide space for those who don't have any. Muggs says she hopes that the concept of a community garden will spread to other areas of the city. "There is a real need for this," she says. "People who don't have space will go out and grow, no matter what."

The Strathcona garden has certainly proved this to be true. There is such a demand for space that the gardeners have begun to create experimental plots on a small piece of unused land adjacent to the garden. Legally this land does not belong to the gardeners, but neither do the three acres of existing gardens.

"Land is inherently political, and land in the heart of an urban area is extremely political."

The five-year lease will come up for its second renewal in 1993. The gardeners are continually working to ensure that there is no excuse for the city not to renew their lease. This means the gardeners try to keep the area as free from garbage as possible and work to gain support from community members who believe in the idea of community gardening. There are some people on the city council who would rather see the land used for a playing field or a city park. It is up to the community to show them that the garden is the best use of the space. This includes lobbying the Parks Board and city hall, holding public events, and generally keeping people informed of the garden activities. Many members of the community are discovering that gardening is political.

Escaping Politics?

Muggs says that when she first began gardening, she told her friends that it was a good way to escape politics. "Now it's turned into the biggest political battle of my life," she says. "Land is inherently political, and land in the heart of an urban area is extremely political."

Although the city owns the land, the terms of the lease make it the responsibility of the gardeners to keep the area free of the garbage others leave behind. Once a month, the Strathcona gardeners head out and pick up used hypodermic needles and condoms and discard the remains of abandoned structures set up by homeless people in the wild-growth section of the garden. This regular garbage collection has reduced the dumping of larger items in the bushes by about 70 percent, but use of the area by drug-users, prostitutes, and the homeless continues. "It's an inner city problem," Muggs says. "We don't like it—but there's not much we can do about it."

Most of the gardeners believe the garden must be open to all residents of the area. The land is legally owned by the city, and some members of the city council see not only the homeless but also the gardeners as squatters. Faced with this, it is easy to see why the gardeners respect the rights of the homeless to be there; after all, they were there first.

After: Caitlin Meggs enjoying the garden.

Photo: Ellie Epp.

The community garden has no fence around it. It is important that people who live in this neighborhood know that they are welcome to come down and help out or just sit on the hand made wooden benches scattered around the garden and orchard. For many, it's the only chance they have to connect with nature and other people. As one gardener said, "The garden grew because pressure came from the community. It wouldn't be there if there wasn't the support." One way the gardeners show their appreciation of the surrounding community is through the surplus-produce table. The community knows that whatever is on this table is free for them to take.

Providing food is not the only way the gardeners connect with the broader community. The gardeners helped to spark a five- neighborhood coalition that successfully killed a city plan to build a high-tech, environmentally unfriendly garbage-processing plant in Strathcona. And their strict policy about chemical-free gardening has extended to the neighboring park, where the gardeners successfully lobbied the community and managed to put an end to the spraying of herbicides on the playing fields. Their example of community organizing was instrumental in stopping herbicide spraying in all Vancouver parks.

The organization and work that is needed to be successful community activists is exhausting. The gardeners are incorporated as a legal, nonprofit society that has an elected board of directors and holds regular monthly meetings. The decisions they make are usually reached by consensus. This is where a lot of personal energy goes into making the garden and its management effective. One gardener says it like this: "So much work has been done, but there is still so much to do. We have to constantly raise money, coordinate activities, deal with city politics, repair and build. I can't wait for the day when we can put all of our energy into dealing with gardening and food issues."

There is a lot more work to be done before that will happen. The gardeners recognize the importance of all aspects of the community garden, from a space where people living in cramped quarters can get physical exercise, to a living example of what community organizing can achieve. Often the basic idea gets lost in the flurry of organizing and shoveling gravel or concrete. Joanne Hochu, a longtime resident and gardener, says, "We think people should move away from buying produce and begin growing in their own backyard. We don't have to depend on foreign markets for our fruit and vegetables. It's not ecologically sound, and the way agriculture is now, every time you buy something you are starting a chain of exploitation."

It's this attempt to move away from harming others and the earth that forms the basis of commitment for these urban gardeners.

Contact:

Antonio M. Austria

Chairman of the Board and President
SPCMBY Fisherfolk Federation and Multi-Purpose Cooperative, Inc.
Dagatan Boulevard, Sampaloc Lake
San Lucas I, San Pablo City
Laguna, Philippines

tel: c/o Rodrigo C. Lachica, Jr.
Philippine Business for Social Progress (PBSP)
tel: (632) 49 70 41 to 52; fax: (632) 48 88 91

17

Fishing for a Voice: Grassroots Organizing among Small Fisherfolk

Antonio M. Austria

What some call "development" is often the destruction of other people's livelihood. For small-scale fisherfolk living near Manila in the Philippines, a proposed tourism-development plan meant that the surrounding communities might lose the areas they had fished for generations as well as the quality of their environment. For those involved in the SPCMBY Fisherfolk Federation, the last decade has been a time of organizing and raising the awareness of people living in and around San Pablo, the City of Seven Lakes. With a growing membership and a stronger voice, they have managed to create a cooperative and become active decision makers in how the lakes are managed.

I am a fisherman and president of the SPCMBY Fisherfolk Federation and Multi-Purpose cooperative in San Pablo,in the province of Laguna in the Philippines. San Pablo is known as the City of Seven Lakes. It is an agricultural center, 87 kilometers southwest of the city of Manila. The seven lakes have a surface area ranging from 14 to 105 hectares. There are four structures commonly found on the lakes: floating fish cages, fish pens, restaurants, and cottages.

I want to share with you my experience as a fisherfolk community organizer.

"SPCMBY" stands for the names of six of the seven lakes in San Pablo that are served by our federation. I started organizing in San Pablo in

Rallying For Change: Antonio Austria addresses a crowd of fisherfolk.
Photo: SPCMBY.

1978 when we, the fisherfolk, were threatened by the city government's plans to use the lakes for tourism.

We realized that if we did not group together to have a stronger voice we could lose our means of livelihood. The proposed tourism plan called for dismantling the structures we use to gather fish for our living—floating fish cages, fish pens and guardhouses—to make way for the construction of hotels, restaurants, floating casinos, and facilities for speedboat racing. We knew these activities would adversely affect the environmental quality of the lakes.

We realized that if we did not group together to have a stronger voice, we could lose our means of livelihood.

I spent many hours, using my own funds going from *barangay* to *barangay* (there are fifteen in the area), talking and holding meetings. Initially, even the bigger fish pen owners shared our concerns and joined us in the cause. It took us almost two years to cover five of the six lakes and to set up organizations.

It was not an easy process. Both upper- and lower-income groups of the lake residents opposed us. As we grew larger, the large fish-pen owners saw us as a threat to their operations, while the small fish-cage and fish-pen operators were afraid to join us due to bad experiences with previous organizing efforts. The bigger, richer fish-pen owners and fish-cage operators and even the municipal government tried to destroy our organization by forming new groups to compete with us.

We registered as a cooperative in 1988. Today the SPCMBY Fisherfolk Federation and Multi-Purpose Cooperative has about six hundred fisherfolk members. Our objective is to improve the standard of living for small fisherfolk. We are working to bring about aquatic reform in the same way that there is agrarian reform.

Being aware of the environmental problems affecting our area, we realized that in order to continue earning our livelihood from fishing, we would have to learn how to manage our resources better.

Environmental Problems in the San Pablo Lakes

Let me give you some idea of the environmental problems affecting the San Pablo lakes.

Overcrowding and Poor Water Circulation

The lack of management and control over development of the lakes in the area has affected the circulation of water and the access of small fisherfolk to common open-fishing grounds. Fish production in both the fish cages and open water has been decreasing. Ninety percent of the fish do not grow on natural feed.

Water Pollution from Sewage Discharge

Untreated sewage and solid wastes are discharged directly into the lakes from residential cottages and commercial structures such as restaurants built along the lakeshore.

Excessive Growth of Water Lilies

Water hyacinth, or water lilies, interfere with fishing and navigating operations. They also damage the net enclosures of fish cages and fish pens during typhoons.

Residual Effect of Fish Feed

Artificial fish feeds, usually in powder and pellet form, are used in intensive fish farming and leave residues that contribute to the deterioration of water quality.

SPCMBY Environmental Activities

Our federation develops and manages programs and services that help achieve our primary objective of preserving the productivity of resources and the ecological soundness of the lake. We work with the Laguna Lake Development Authority (LLDA), the government body responsible for the development and management of the seven lakes and

Laguna de Bay, the country's largest fresh-water resource. For example, SPCMBY reviews all requests to construct fish cages, fish pens, restaurants, and cottages. We also have been able to suggest regulations to be implemented by the LLDA, such as those governing the operation of fish pens and fish cages.

We are testing and evaluating new technologies, such as the floating hatchery. Some of the techniques we have developed have been adopted by the government's Bureau of Fisheries. We have developed a floating fish-cage production program. The high profitability of floating fish cages has led to overcrowding and consequent lower productivity. Our program regulates the operation of fish cages and ensures the preservation and productivity of our main source of living.

We are involved in local advocacy by fighting against the city

The (Un)making of a Fisheries Law

Tess H. Lingan and Tony Debuque

The following discussion of the predicament faced by subsistence fisherfolk in the Philippines is part of a larger article published in the January February 1991 issue of Phildhrra Notes, *a newsletter produced by the Philippine Partnership for the Development of Human Resources. The article describes how fisherfolk, whose livelihoods are threatened by larger-scale commercial fishing operations and other forms of development, are organizing and pressuring the government for aquatic reform. Fisherfolk have formed a nationwide coalition to effect change in legislation that they hope will keep the waters open for fishing at a small and sustainable level. This article provides background on the situation; but exactly how some of these fisherfolk are organizing is revealed in the personal story of Antonio Austria and the SPCMBY Fisherfolk Federation.*

Except perhaps for the usual images associated with the fishing trade, little is known of the small fisherfolk. In the Philippines, in particular, not many people know that this sector alone feeds over nine million Filipinos, provides 77 percent of the population's protein requirements, and contributes almost half of the nation's total fish catch.

Best characterized in distinction from operators of commercial fishing vessels and fish ponds, small fisherfolk are generally those who use anything from boats that run on 16-horsepower engines to the crudest of fishing gear. They fish in "municipal waters," which include streams, lakes, tidal waters, and the part of the sea that is within three nautical miles, or seven kilometers from the shoreline.

government's lakeside tourism-development plan. And since 1987 we have also been involved in advocacy at the national level, as a result of our membership in the Nationwide Coalition of Fisherfolk for Aquatic Reform (NACFAR). We were involved in drafting and lobbying for the passage of the Comprehensive Fisheries Reform Code of 1990 (or the Unity Bill). This legislation seeks to protect the country's fishery and aquatic resources from exploitation by foreigners and is committed to the conservation, management, and development of these resources.

For almost ten years we depended mainly on our own resources for organizing. Then nongovernmental organizations assisted us. This has permitted us to get external funding to support our different programs and to broaden our network by linking with other fisherfolk organizations such as NACFAR. But more important, our links with NGOs have helped us in institutional development through, for example, research

In 1987 government statistics placed the number of small fisherfolk at a little over 500,000. In 1990, fisherfolk leaders estimated their numbers to be between two and four million, while Senator Agapito Aquino reported that more than eight million people engage in small-scale fishing as a means of livelihood. It is difficult to keep track, they say, because fishing is not a job you take tests or register for. Hitch a ride to the sea (if you can't get your own *banca* or boat), bait a line, and you're in business.

But perhaps another reason for their low profile is that they are one of the least-organized sectors in Philippine society. Of possibly eight million small fisherfolk in the country, only 200,000 are organized, and these are concentrated in the Visayas. For the most part, fisherfolk have been a silent sector, and this characteristic has been attributed to the solitary nature of their work.

Yet silence in their case does not imply the absence of problems. Like deep waters whose unruffled surface often belies turbulence beneath, Filipino fisherfolk are demanding reform as urgently as the ranks of the peasantry. Their issues are similar—the unsustainability of present methods of resource use, the emphasis on maximum rather than optimum resource yields, export-oriented and import-dependent national policies, undernourished food producers, low real income, low "farm gate" prices, the absence of producer control over the marketing of products, the lack of modern equipment and technology, and a law that favors commercial over subsistence operations.

But while both are food producers, the two sectors differ in the nature of production relations. While farmers are directly related to landlords in the production process, fisherfolk relate only to the sea and nature. Philippine territorial waters are owned by no one. They are not subject to private ownership and so cannot be parceled or titled out. Given the length of the coastline, each and formal training in federation development, management and team building. Through the Philippine Development Assistance Program (PDAP) we have received help from a Canadian NGO to purchase fish food and materials for the construction and repair of fish pens and fish cages.

The SPCMBY federation is a partner organization of both government and nongovernment groups. We have proven this. Previously, we only followed programmes, policies, and regulations made by the government. Now we help plan our own programs, we are recognized nationally, and some of our proposals have reached the Congress. In 1989 we were able to borrow for the first time from the Land Bank of the Philippines.

small fisherfolk should be able to operate in at least one square kilometer of fishing ground, but does this really happen?

The Philippine coastline, all 34,600 kilometers of it, is still the longest in the world, and its inland waters are still where they should be, but current patterns of resource use have drastically shrunk the small fisherfolk's world in recent years. Ecologically destructive fishing methods are depleting the marine and fishery resources at nonreplenishable levels. The mangroves that serve as spawning and breeding grounds have all but disappeared in recent years. The coral reef, in which sea life makes its home and so ensures the presence of fish and other aquatic life in territorial waters, has been either destroyed, ravaged or sold abroad. The rivers and lakes, on the other hand, are dying from industrial, agricultural, and household waste.

The lack of sophisticated equipment and technology limits eight million small fisherfolk to municipal waters, which are already overfished. Venturing farther offshore, however, entails higher fuel costs, which they cannot afford, and greater risks to their personal safety. Furthermore, owners of commercial fishing vessels have even made inroads into municipal fishing, while maintaining their dominance in offshore and deep-sea fishing operations.

And as if the small fisherfolk's world isn't already crowded, fish pens, fish cages, and fish belts have sprung up in recent years right in their beleaguered turf. Besides rounding up whatever is left of the aquatic life in the area, these structures, which ignore the all-too-obvious fact that a confined space can only accommodate a limited number of fish, are causing fish to die for lack of air and food. This has reduced small fisherfolk to scavenging for whatever the pens and cages have not fenced in. Not a few cases of harassment and outright murder of fisherfolk have been committed by the guards of these "preserves," all on the charge that the fisherfolk are "Trespassing." And we thought no one owned the sea ....

18

A Community of Intention: Huehuecoyotl

Alberto Ruz Buenfil

It can be fate or coincidence that brings people together in community. But there are also many communities that form by intention. Intentional Communities is a term often used to describe these groups of people who, for whatever reasons, have decided to share their lives together. The values that generate a communal lifestyle very often include a deep concern for the well-being of other people and the environment.

Alberto Ruz Buenfil is one of the founding members of Huehuecoyotl, a small village located southeast of Mexico City. In the following article he describes how the group evolved from a traveling theater to a settled community nestled in the dry, dusty hills of Mexico.

Huehuecoyotl, the "place of the old, old Coyote" in the Nahuatl language, is an alternative village located in the Tepozteco mountain range and in the Cuahunahuac bioregion in the state of Morelos, Mexico. We are a group of twenty adults and eight children from different nationalities and backgrounds, who founded the community in 1982. However, our origins go back to the early 1970s, when we first came together in southern Sweden at the Situationnist Bauhaus of Drakabyget, near Orkelljunga.

For more than ten years, our community grew in size and form as we traveled and used theater as a vehicle to convey information across political and cultural borders. We performed for indigenous and marginalized communities in the small villages and schools, streets and plazas, cultural centers and universities we passed along the way. This long odyssey through cultures, peoples, and ethnic and social diversity

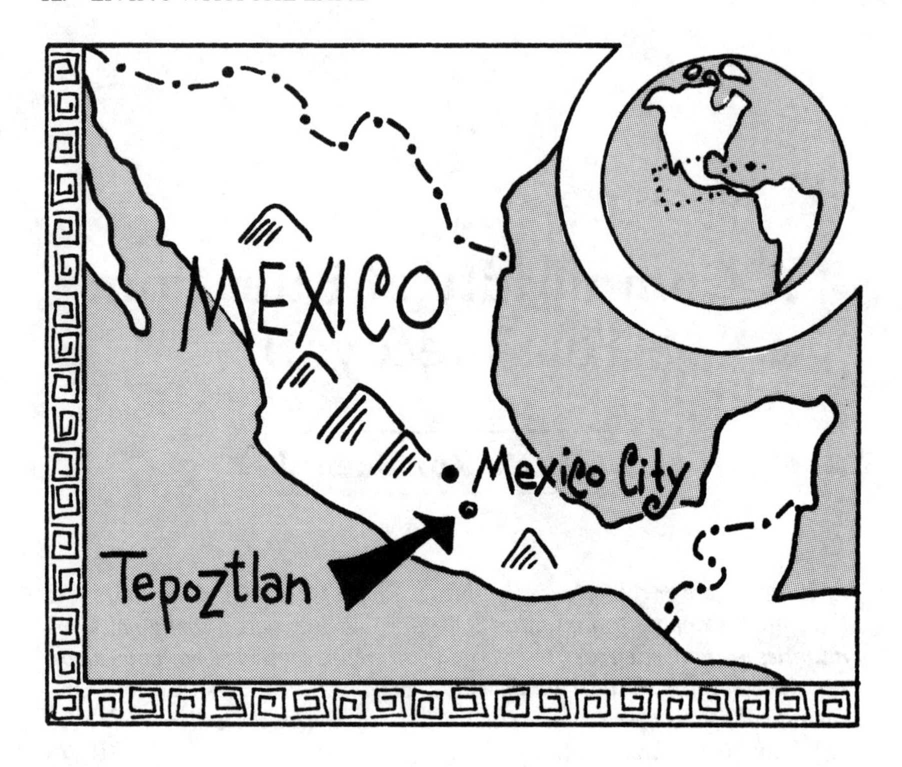

Contact:

Alberto Ruz Buenfil

Huehuecoyotl
A.C. Experimental Workshops Center
A.P. 111 Tepoztlan
Morelos, Mexico

set the basis for our collective education. It also took us on an inner journey, during which our original goal of changing the world evolved into recognizing the need to change our own lives as well as the external environment.

As our children grew older and our community larger, we found that we needed to have a home base. We looked at many factors—political, cultural, economic, physical, and spiritual—and eventually decided that Mexico would be the place for our permanent home. Almost no other alternative community existed in this part of the world. After about half a year of searching the four directions, we put down our roots in Huehuecoyotl. It is near Tepoztlan, a very magical spot where artists, seekers, healers, and crafts-peoople have chosen to live.

This long odyssey through cultures, peoples, and ethnic and social diversity set the basis for our collective education.

We eventually acquired a small piece of land—an earth bay of five acres of flat land in terraces, surrounded on three sides by the vertical slopes of the mountains and facing the east on its fourth, open side. The terraces were lined by old fruit trees: avocado, mexican cherry (*capulines*), peaches, *chirimoyas*, and *zapote blanco*. Oak, manzanito, and rare pines are found on top of the mountains, and low brush covers the volcanic rivers that limit the property.

We bought the land from a local Nahuatl farmer, and it's considered communal land, belonging to the joint owners from the village of Santo Domingo Ocotitlan. The property is considered a trust and nobody can sell without the full agreement of all twenty members of the community. After five years of trial, we and our new village of Huehuecoyotl were finally accepted by members of the older village, resulting in both responsibilities and duties toward the community as a whole.

These responsibilities take many forms, including a popular clinic for neighboring villagers and a women's project working to solve perhaps the biggest problem in the area—water conservation. We are also completing construction of a school and a cultural and ecological documentary center for all the neighboring villages around Tepoztlan.

For the last two years the school, called Cetiliztli, has played an active role in raising people's awareness about environmental and ecological problems. Our first goal with the school was to provide a more regular education for the children of Huehuecoyotl. The school has now grown

The Place of The Old, Old Coyote: Meeting in the shade of the Amate tree in Huehuecoyotl.

Photo: Alberto Ruz Buenfil.

into a center of alternative learning for the surrounding communities. Children and teachers are involved in cleaning rivers, creeks, and streets, educating people on recycling; organizing public parades on Earth Day; and fighting the use of pesticides, the destruction of forests, lakes, and local mountains, and the impact of Sunday tourism from Mexico City on the villages and native culture.

Education for our children was one of the main reasons we decided to find a home base nine years ago. The children were growing older and were asking for more permanent relationships with other children and with their educational process as well. For many years our children attended different schools as we traveled, or else they learned on the road with some of the parents who had training or an interest in teaching. Largely the children were educated through traveling and meeting people from different cultures.

When we settled near Tepoztlan we already knew many people in the area who were living alternative lifestyles. Together we created our first *escuelita* (little school), to give local children a different kind of education. That was in 1984. Through the years it has developed and changed form, location, teachers, and systems. Only two years ago it was officially recognized by the Ministry of Education and until then it was only with the support of a local native elementary school that our children got their the support of a local native elementary school that our children got their papers with which to further their studies after completing the sixth year of elementary education.

Through the years, the school grew from ten children to more than fifty, serving not only the children of our friends but also children and teachers from native villages. Today the school is part of a larger, wholistic approach to working ecologically with the surrounding communities. The school is still growing, along with the clinic, ecology center and the Luna Nueva women's project, which is building water cisterns in dozens of houses in Santo Domingo and Amatlan.

Our involvement with the surrounding communities required that we first establish our homes in Huehuecoyotl. The next steps we took as part of the new community were to open an access road to the land and to build a dam at the foot of the waterfall for collecting water in the rainy season. We are at an elevation of 5400 feet and for six months of the year it is very dry and dusty.

We constructed our first homes with local material using adobe (earth bricks) for the walls and lava stones for foundations and the lower walls. The houses have wooden beams and the roofs are covered by tiles, wood, or ferro-cement. Now there are thirteen houses, each with its own cistern to hold the yearly allocation of water and a gray-water recycling system; some even have Vietnamese-style, dry outdoor latrines. The construction styles are as diverse as the cultural influences, but there is a unity of harmonious, simple, and beautiful shapes, materials, and proportions. Most of the houses are half covered by plants, trees, and flowers.

We are all active in different communal workshops, producing honey, ethnic musical instruments, natural aloe vera products, and edible mushrooms. Together we produce a modest amount of food in the community—vegetables, eggs, honey, fruit marmalades, and bread. There's a carpenter shop, a video and audiovisual center, three performing groups (which have produced four audio cassettes) and several people involved in astrology, education, and healing. We develop some of these activities at the school.

Most of us make our living partially on the land, while others leave for three or four months a year to work in the United States or Europe, raising capital for tools, instruments, and the construction of houses and communal projects. Our economies are individual or familial; each person is responsible for him or herself and resolves financial problems as desired. We do have a communal account for minor expenses and for improvements on the land like roads, dams, gardens, and communal houses.

We make all decisions in monthly councils on the first Sunday of every month. We have an agenda and discuss the most important problems concerning Huehuecoyotl. We make decisions by consensus since we do not want to rely on voting, which can leave individuals and minorities out of an all-group process.

Huehuecoyotl depends on its own resources and has never received financial support from any institution, national or international. Self-reliance has been the basis of our ideology, and what we have accomplished in our nine years is entirely a result of the work of our members. This has also granted us full independence, freedom, and an uncompromising attitude regarding institutions of all political and economic systems.

The main problem we are facing these days is the seeming lack of interest in communal projects, as people tend more to do their own thing rather than get involved in collective endeavors. We hope this is a temporary stage and believe the lack of a functional communal space is one explanation for this problem. There is an expectation that the communal building will allow more collective activities in the near future. We are also counting on the use of this space to resolve the problem of yearly migration to the north for money, instead reserving such travel for communication, performances, and the expansion of networks.

The most important lesson Huehuecoyotl can share with others is the possibility that a group of people with different backgrounds and experiences, even cultural and national differences can experiment and create an alternative way of living for themselves and their children. Huehuecoyotl serves as a living example of self-reliance for people from the four directions. We think the quality of life here is incomparable, and there is no one here who thinks of living somewhere else.

In spite of daily difficulties and small crises, life flows easily, children are growing healthy and happy, and grownups are learning to become wise elders with the passing of the seasons. In the future, we envision our alternative village expanding and including more families, their children and elderly parents, a few more workshops, and a storefront.

Huehuecoyotl has already inspired many people to begin collective projects, serving as an example to the whole environmental movement in Mexico and abroad. We are a living and working alternative, an example of the process of permanent growth. We are turning slowly into a new model for small-scale, sustainable communities for the coming millennium.